Vashon Island's Agricultural Roots

Cover photo:
Mitsuno Matsuda harvests the bounty of her family's strawberry field on Vashon Island.
Photo courtesy of her daughter, Mary (Matsuda) Gruenewald, who is interviewed in Chapter 6. Digital photo restoration by Imago Imaging, Seattle.

Vashon Island's Agricultural Roots

Tales of the Tilth as Told by Island Farmers

Pamela J. Woodroffe

Writers Club Press
San Jose   New York   Lincoln   Shanghai

Vashon Island's Agricultural Roots
Tales of the Tilth as Told by Island Farmers

All Rights Reserved © 2002 by Pamela Joan Woodroffe

No part of this book may be reproduced or transmitted in any form or by any means, graphic, electronic, or mechanical, including photocopying, recording, taping, or by any information storage retrieval system, without the permission in writing from the publisher.

Writers Club Press
an imprint of iUniverse, Inc.

For information address:
iUniverse, Inc.
5220 S. 16th St., Suite 200
Lincoln, NE 68512
www.iuniverse.com

ISBN: 0-595-24133-6

Printed in the United States of America

Funded by a grant from the King County Landmarks and Heritage Commission and funds from the Vashon-Maury Island Land Trust. Complete tapes and transcripts of the 300 pages of interviews, called the Roots Project, may be found at the King County Library system, Vashon Island branch, 17210 Vashon Highway SW, Vashon, WA 98070, phone (206) 463-2069.

King County
Landmarks & Heritage Commission
Hotel/Motel Tax Fund

VASHON
MAURY
ISLAND
LAND
TRUST

Pamela J. Woodroffe, Project Director
P.O. Box 17658
Seattle, WA 98107
pamwood@earthlink.net
www.pamelawrites.net

Foreword

The Vashon–Maury Island Land Trust works to conserve natural ecosystems and the rural character of our islands. Over the past decade, we've focused our efforts primarily on wildlife habitat—permanently protecting nearly 350 acres of key land. The Shinglemill Salmon Preserve and Christensen Pond Bird Preserve are two successful efforts to protect natural areas.

However, it is because of the islands' rural character that many of us have chosen to live here. "Rural character" has always been a thorny term for islanders to define. It certainly means open space, forests, streams, and pastures. However, it also refers in some way to agriculture and making a living off the land.

This book grew out of "Tales of the Tractor," an educational series offered in partnership with the Vashon Island Growers Association. We're proud to sponsor this book as an important link to the agricultural history and rural character of Vashon and Maury Islands. This view to the past keeps our sense of community strong and reminds of us of our legacy for generations to come.

Because the land trust's focus is healthy streams and forests for salmon and birds, we're especially appreciative of today's farmers who share a compatible mission. Growers who don't use chemicals, keep pastures in open space, and feed island families with homegrown produce all contribute to land preservation and stewardship. This and, most importantly, what we each do with our own land will shape the islands' future.

Board of Directors
Vashon–Maury Island Land Trust
2002

PREFACE

This book was born three years ago when Dave Warren, former director of the Vashon–Maury Island Land Trust, asked if I'd like to write a book about Vashon's farming history and future. I thought it would be a good opportunity to combine my background in writing and in farming—I'm the former owner of Sweet Woodroffe Herb Farm on Vashon Island.

Gathering these stories about Vashon's rich agricultural history has been an honor. These farmers' and growers' generosity of spirit, kindness, and welcome are truly inspirational. Their love of the land and their undaunting, indefatigable efforts to make their living from it can be seen through the conviction of their words, the intermittent tears in their eyes, the calluses on their hands, as well as the fruits of their labors.

Historian Bruce Haulman gives us an overview of Vashon's agriculture, which documents some of the cycles experienced by farmers here. In Part I, *Homesteads, Fishing, Fieldwork, and Floriculture,* longtime residents and farmworkers give us their personal view of the history of the islands' agricultural development. Bill Mann tells us 60 years of tales from the tractor. Eugene and Peggy Sherman share their recollections of the island's strawberry farms, chicken farms, and myriad other crops. Tom Lorentzen's account tells about the Norwegian fishing and farming settlement in Cove and Colvos. Tom Beall Jr. recollects his family's 100 years of greenhouse-grown roses and orchids, with a brief history of the Beall Greenhouse Company.

Part II, *Strawberry Fields Forever,* looks at the hard-work ethics and experiences of World War II and internment of some of Vashon's Japanese-American strawberry farmers, August Takatsuka and Mary Matsuda Gruenewald. The story of Masahiro Mukai and the Vashon Island Packing Company was compiled by historian Mary J. Matthews.

Part III, *Tree Fruit Orchards Branch Out,* tells about more-recent farmers here from the 1950s on, including Robert and Betsy Wax Sestrap's chicken farming, fruit orchards, and specialty food products of Wax Orchards Farm. Pete Svinth, of Pete's Family Trees, tells about the spirit of growing trees.

In Part IV, *Community-Supported Agriculture and Contemporary Farmers,* we hear from present-day farmers, growers, and food producers. Bob and Bonnie Gregson, of Island Meadow Farm, talk about subscription farming and apprenticeship programs. Ron Irvine, of Vashon Winery, tells of his twenty-five years of wine and cider production. Jasper Koster and Martin Nyberg, of Green Man Farm, are enthusiastic and passionate about the urgent need for healthy food. Amy and Joseph Bogard, of Hogsback Farm, talk about teaching kids the value of farming. Margaret Hoeffel with Jean Bosch of Blue Moon Farm, tells us about specializing in winter greens and cooking classes; Margaret's husband, Martin Koenig, describes raising money to enable the Vashon Island Growers Association to buy Village Green for farmers and the Vashon community.

A project naturally takes on the bias of the director, so I'll admit I wanted to know whether farmers can make a good living off the land. How do they balance their farm with their family life? Would farmers of the past do it all over again? What marketing, business planning, or crops or products make farming viable? What farm subsidy programs and political policies toward farming are effective, and which ones make matters worse? What is needed to support sustainable agriculture for Vashon—and, indeed, the country's—future? What can citizens and consumers do to help bring healthy food to their family's dinner table?

The "Roots project," this book's working title, is only the beginning of a description of farm life. I have distilled each of the ten-to thirty-page interviews into all-too-brief vignettes. The entire recorded tapes and the transcripts of the interviews are available for review in the King County Library, Vashon branch.

I have tried to preserve the narrative voice and intent of all of the farmers and the accuracy of their words, as well as fact-check them against the backdrop of local and global history. I welcome comments and suggestions, and hope that other historians will build on this project to document more fully the history of Vashon's agriculture.

Acknowledgments

Thanks to the grant-writing mentorship of Mary J. Matthews, of Island Landmarks, and Dana Ilo, of the Vashon–Maury Island Land Trust, the King County Landmarks and Heritage Commission awarded a Special Projects grant to support this project.

Also of help were endorsements by Bob Gregson, Island Meadow farmer and member of the King County Agriculture Commission; Tom Lorentzen, current and past Vashon farmer and farmworker; David H. Swain, Vashon–Maury Island Heritage Association; and Sharon Price, Island Landmarks and the Committee to Preserve the Mukai Farm and Garden. The project continued with shepherding and support by Julie Burman, current executive director of the Vashon–Maury Island Land Trust, and Holly Taylor, Heritage Program coordinator for King County Landmarks and Heritage Program; as well as encouragement by Priscilla Beard, Vashon–Maury Island Land Trust. Contributions and historical information were provided by Mary J. Matthews, Island Landmarks; Bruce Haulman, historian; Mary Jo Barrentine, Vashon–Maury Island Heritage Association; and Rayna Holtz, Vashon library. Historical photographs were generously provided by Vashon–Maury Island Heritage Association, Arthur (Tom) Lorentzen, and Mary Matsuda Gruenewald. Contemporary photographs of Vashon's farmers and farms were generously donated by Kathleen Webster and Ray Pfortner, who captured the heart and soul of the farmers and their land, respectively, only long enough so it could be released to the public.

Thanks to Rochelle Wallace, Anacortes, who loaned her transcribing machine for the duration of the project. To Kris Fulsaas, editor extraordinaire, who supplied me with the most respectful, kind, and thorough editing process I have experienced in twenty-fiveyears as a published writer. And to Dion Cantellay, Seattle Image Setting, for the top-quality scans of the photos of contemporary farmers.

Credit is also due to my daughter, Angie Woodroffe Weiss, who quietly wrote her own books, patient with a mother who was immersed in a world of strawberry weevils, rose hybridization, biodynamic compost experiments, internment camps, chickens choking on roofing nails, and Holy Grail cider quests.

And thanks to the farmers, who were kind enough to share their hearts and souls and their love of farming through their stories; this book is a tribute these humble heroes, their efforts and their works.

For my daughter, Angie Woodroffe Weiss, with high hopes of an ever-plentiful, wholesome, healthy food supply for her and the generations that follow

Contents

Foreword vii
By Vashon–Maury Island Land Trust

Preface ix

Acknowledgments xiii

Introduction: Vashon's Agricultural Past xxiii
By Bruce Haulman
Historian Bruce Haulman describes the cycles of Vashon–Maury Islands' farming enterprises beginning with the S'Homamish people, followed by loggers and small-scale farmers, then large-scale poultry, berry, and greenhouse ventures, culminating in present-day sustainable-agriculture farms. Vashon growers and farmers have been national prize-winners in chicken-rearing, the largest regional producers of greenhouse roses, and top producers of tons of juicy, red strawberries, and have led the standards in tree fruit science. Today's contemporary farmers make their living through niche marketing and community-supported agriculture.

Part I: *Homesteads, Fishing, Fieldwork, and Floriculture* 1
Stories by Longtime Farmers and Agricultural Workers

Chapter 1: William Mann: Tales from the Tractor 3
During more than sixty years of tractor work on Vashon Island, Bill Mann has cleared land for growers and, more recently, developers. He is known for the reindeer-shaped pond he built.

Chapter 2: Eugene and Peggy Sherman: Family Farms 9
The Shermans recall the days when Vashon was largely strawberry and chicken farms, and recount all of the myriad crops that have been grown here, including garlic, ginseng, and daffodils.

Chapter 3: Arthur Tom Lorentzen: Norwegian Settlement at Cove and Colvos 14
With his stories about the small Norwegian farming and fishing settlement near Cove and Colvos, Tom Lorentzen regales with his recollections of working on the fruit farms on Vashon.

Chapter 4: Tom Beall Jr.: The Beall Greenhouse Company 25
Tom Beall Jr. recalls his three-generation family rose-and orchid-growing business and the multiple greenhouses that housed the plants.

    History of the Beall Greenhouse Company
    Compiled by Mary J. Matthews

    In the 1970s, fuel prices rose and competition from other countries forced the closure of what had been Vashon Island's largest employer.

Part II: *Strawberry Fields Forever* 39
Vashon's Strawberry Industry and Japanese-American Farmers

Chapter 5: August "Augie" Takatsuka: From Strawberries to Christmas Trees 41
Augie Takatsuka, one of the large-scale Japanese-American strawberry farmers of Vashon, talks about his crops being harvested by Native Americans and other migrant workers, his internment during World War II, the subsequent loss of part of his foot when he fought for the United States in World War II, and his Christmas tree and wreath enterprise.

    A Brief Time Line of World War II

Pamela J. Woodroffe   xix

Chapter 6: Mary Matsuda Gruenewald: The Strength of Family 53
Mary Matsuda Gruenewald recalls her family's work ethic and how they provided instruction and peppermint candy incentives to kids who helped harvest their crops.

    A Personal Account of the Internment of Japanese Americans

    Mary Matsuda Gruenewald gives an astounding account of her internment experiences as a seventeen-year-old, and talks about the power of forgiveness; she is writing a book about this experience.

    U.S. Laws Affecting Japanese and Japanese Americans
    Compiled by Mary J. Matthews

Chapter 7: Masahiro Mukai's Vashon Island Packing Company: The Vashon Island Strawberry Industry 69
By Mary J. Matthews
VIPCO, an innovative berry freezing plant on Vashon Island owned by Masahiro Mukai, changed the strawberry-growing industry on Vashon.

Part III: *Tree Fruit Orchards Branch Out* 77
Midcentury Food Producers from the 1950s On

Chapter 8: Betsy and Robert Sestrap: Wax Orchards Farm 79
Betsy Wax Sestrap tells about her experiences growing up on Wax Orchards Farm, which first raised chickens; later she grew tree fruits with her husband, Robert. They moved into the preserves and fruit products business, specializing in fruit-sugar-sweetened products, and maintained a small airport.

Chapter 9: Pete Svinth: Pete's Farm 86
Pete calls himself a grower, not a farmer. He recalls his years grafting fruit trees, having an honor-system produce stand, and his pumpkin patch enterprise. He rediscovered his Native American roots and moved off Vashon to live closer to his tribal community and help them farm.

Part IV: *Community-Supported Agriculture and
Contemporary Farmers* 93
Providing Healthy Food and Hope for a Future with Sustainable Agriculture

Chapter 10: Bob and Bonnie Gregson:
Island Meadow Farm 95
Bob and Bonnie Gregson were farmers of Island Meadow Farm for about ten years, and published a book about their small family farm. They marketed their products through subscription customers and their produce stand. Recently they retired from farming and negotiated an ongoing apprenticeship program that keeps part of their land in farming, so Bob can stay active in King County farming issues.

Chapter 11: Ron Irvine: Vashon Winery 102
Ron Irvine tells about grape growing and apple growing at Vashon Winery, and the moment on a picnic in London where he tasted a cider that prompted a lifelong passion for wine and cider production. A former co-owner of a retail wine shop in Seattle, he published a book about Washington state's wine industry.

Chapter 12: Jasper Koster and Martin Nyberg:
Green Man Farm 107
Jasper Koster and Martin Nyberg of Green Man Farm have an organic subscription farm and a small produce stand, and also sell at the Vashon Farmers Market. They talk about the travails of a couple who mix city commuting and home farming, and about their political activities supporting sustainable agriculture through the Vashon Island Growers Association and other organizations.

Chapter 13: Amy and Joseph Bogard: Hogsback Farm 114
Amy and Joseph Bogard bought part of Tok Otsuka's former strawberry farm, which still has some of the migrant cabins standing, where they homesteaded while they built their house. Like Jasper Koster and Martin Nyberg, they farm organically and one of them commutes to a job in the city. They like to host schoolkids' visits to Hogsback Farm for education and fun.

Chapter 14: Margaret Hoeffel with Jean Bosch:
Blue Moon Farm 119
Margaret Hoeffel partners with Jean Bosch growing greens and other crops on Blue Moon Farm at a cohousing site on Vashon. Margaret spearheaded the "Tales from the Tractor" public education series when she worked for the Vashon–Maury Island Land Trust, which prompted the development of this book.

Chapter 15: Martin Koenig: The Raising of Village Green 124
Martin Koenig raised funds to publicly acquire Village Green, the downtown Vashon land that houses Vashon Island Growers Association's produce stand, known as the Vashon Farmers Market.

Afterword 129

Appendix: Vashon–Maury Island Farming, Historical, and Land
Preservation Organizations 131

Index 133

Contributors 141

About the Author 143

Introduction

Vashon's Agricultural Past
By Bruce Haulman

Writing about the agricultural history of Vashon–Maury Island is a bit like writing the regional history of a section of the United States. There is a larger story to be told, but because of the constraints of focusing on a particular region, much of the larger story must be ignored or relegated to secondary importance. So it is with the agricultural history of Vashon. Agriculture forms only a part of the larger story of the island, and yet, as with a region of the United States, the part it plays is essential to understanding the story of the whole.

Vashon–Maury Island's history began as the retreating ice sheets of the Vashon Glacier exposed the landscape about 15,000 years ago. The human history of the island began when Paleolithic nomadic hunters crossed the island about 12,000 years ago. The island was then inhabited, beginning 3,000 to 8,000 years ago, by Native peoples who called themselves the S'Homamish. Vashon was "discovered" by Europeans in 1792, "conquered" by Americans in the Puget Sound Indian War of 1854, "settled" by a diverse group of Euro-Asian-Americans in a series of separate communities linked by only water in the 1880s and 1890s, and "developed" into the unified community it is today during the second half of the twentieth century.

Agriculture on Vashon–Maury Island began early, perhaps as long ago as 3,500 years, and today still plays an important role in the island's process of self-identification. The history of agriculture on the island is divided into five distinct phases, each one focusing on different "crops," methods, and markets.

The first agriculture on Vashon–Maury Island developed as early as 3,500 years ago as the riverine Marpole Culture evolved into the Coastal Salish Culture of the S'Homamish people who inhabited the island. Although largely dependent on the marine resources of the island, the S'Homamish people also cultivated the uplands of the islands using burning regimes to clear the land, maintaining swales for their horses, allowing root plants to flourish for harvesting, and encouraging berries to flourish.

The second stage of agriculture on the island emerged with the arrival of Euro-Americans beginning in the 1860s and was dominated by small-scale logging and subsistence farming. This stage lasted for nearly thirty years and established the basic patterns for settlement and development of the island.

The third stage of island agriculture saw the emergence of large-scale industrial logging, the development of substantial greenhouse operations, and the expansion of farms specializing in diverse products such as poultry, dairy, vegetables, and berries. Beginning in the 1890s, this stage of agriculture dominated the island economy until the agricultural depression of the 1920s and the Great Depression of the 1930s led to an agricultural collapse and a decline in island population.

World War II revived small-scale island agriculture into the fourth stage. The influx of wartime workers during the war, a new professional commuter population in the 1950s, and a back-to-the-earth counterculture population in the 1960s and '70s reestablished the small farm on the island as either a part-time operation supported by outside income, or a subsistence operation to support an alternative lifestyle.

Beginning in the 1980s and developing strongly in the 1990s, the current fifth stage of island agricultural history began with the establishment of a number of sustainable-agriculture farms, a revival of organic farming methods, and emerging markets for locally grown produce.

Pamela Woodroffe's fine collection of stories and reminiscences by Vashon–Maury Island farmers adds significantly to our understanding of the important role agriculture has played in the island's past. This collection also highlights the role agriculture continues to play in Vashon–Maury Island's continual process of self-identification. Agriculture is the symbolic and mythic foundation on which the island's self-identify is built, even though that foundation is less and less apparent in the island's modern service-and commuter-based economy.

Part I

HOMESTEADS, FISHING, FIELDWORK, AND FLORICULTURE

Stories by Longtime Farmers and Agricultural Workers

Chapter 1

William Mann: Tales from the Tractor

William Mann has tilled Vashon's land for more than 60 years.
Photograph © Kathleen Webster

For more than fifty years, William John Mann and his brother Harold Mann have tilled, disked, baled hay, and cleared Scotch broom, blackberries, and brush on Vashon—returning each year to bring fertility to several thousand acres of the island's agricultural landmass. More than anyone, they know the layers of history of the landscape here; its stories are told through its tilth. They're known by the locals as the Mann Brothers—the ones who dug the giant reindeer-shaped pond on their property that each winter is outlined in Christmas lights.

Bill's home on Cemetery Road is more like a bachelors' garage, with a couple of old easy chairs and a woodstove. Bill nods to introduce his brother Harold, who has shuffled in for a well-earned coffee break. Bill, a careful and deliberate man, is slow to warm up to the prospect of telling his "tales from the tractor," but when asked about his customers, he gets a twinkle in his eyes, sometimes a tear (he's outlived most of them), and a half century of island history comes pouring out—decade by decade.

In the late 1930s, says Bill, "The whole island was one big berry and chicken farm. All along the main highway up, there were solid farms, all the way from Morgan Hill clear up to Jack's corner [by the Heights' water tank], on both sides of the road. Those were a lot of 5-and 10-acre farms," Bill recalls; "everything from the [Seattle's Best] coffee shop and back to the Bone Factory [Pacific Research] was 5-to 20-acre pieces. That whole thing was basically solid currants. Then right there at 188th," he continues, "Masa Mukai cleared that 40-acre piece, and that was all strawberries when he built the cannery. He owned 40 acres south of the cannery, and had the 60 acres where the Catholic church is.

"You used to be able to see clear across the island. From Vashon clear over to the Catholic church area, you could look out and see all the farms. Now you can't see because of the brush that's all in there. That whole big block in there, between 107th and Vashon Highway and Cove Road–Bank Road was just all solid berry farms. Right in the town of Vashon, right where the Dairy Queen sits now, there used to be a big farmhouse. That was all berry fields, clear back down around where the Vashon Post Office is—just solid berry fields there," Bill remembers. "The whole middle section of the island was virtually totally cleared."

He and his brother grew up in farming and learned about dairy, chickens, and land cultivation from their parents, Karl and Rosie Mann. The family moved to Vashon in 1938 and lived on Bank Road for six years, where they raised 200 laying hens. Like many chicken farmers, they sold the eggs to a broker, Royce Wise, who picked up the eggs from Vashon farms and *candled* them—graded and sorted them. Bill estimates that there were several hundred chicken farms on Vashon in the 1940s. Many farmers—particularly currant and cane berry growers—also raised chickens for the fertilizer they'd produce. However, chicken manure created too much foliage growth to be used successfully with strawberries, Bill says.

In 1944 the Mann family moved to Cemetery Road, where they raised some milk cows for a while, then turned their focus to tractor work. Bill quit before his last year of high school and took a job driving a milk truck for a local dairy farm. The family grew 3.5 acres of boysenberries and raspberries that they sold to a cannery in Puyallup. During that time, the cannery sent trucks to the edges of the berry fields to collect the fruit grown on the island. As farming became less profitable, the trucks quit coming to the island, and the fruit was collected at a loading dock where Owen's Antiques sits today.

In the 1940s, when World War II broke out, a lot of the farmers went to work in the defense plants, Bill recalls. Few returned to farming. The Japanese farmers were sent away to internment camps (see Part II, Strawberry Fields Forever); after the war, Yoneichi Matsuda (see Chapter 6), Augie Takatsuka (see Chapter 5), Tok Otsuka, Frank and Jimmy Matsumoto, and Tashio Fujioka returned to farming on Vashon for several years. During the war, gasoline was rationed and the Manns had to sign up for fuel to keep their business going.

In the 1940s, there were only three other people doing heavy tractor work; when the Manns retired, they were the only ones with the equipment. They were perfectly positioned to fill the demand for land cultivation work. When the region suffered recessions and Boeing layoffs, the Manns kept in business, plowing small family gardens for out-of-work families seeking to save money on their food expenses.

No matter what the need—pestilence, market changes, economic challenges—the Manns' skills were always in demand. In the 1950s, currant farmers were plagued with flies that stung the fruit, and hundreds of acres needed to be plowed under. Later, after a chemical dust was discovered to combat the flies, currants were seen as a lucrative crop, so the Manns plowed fields to cultivate them. When there was talk of a bridge being built between Vashon and the mainland, many people subdivided their properties and built houses, anticipating a boom. The Manns were there to clear the land. But the bridge was never built.

The 1950s brought a dwindling labor force, with changes in the labor laws as farmworkers sought better-paying jobs. More chemicals were used on the fields to make cultivation and growing more efficient, to offset the loss of hand labor.

"You used to be able to make more profit with the hoe, and chemicals are expensive," Bill says. "But what else do you do? When you've got acres and acres of strawberries, you've gotta have a lot of people or chemicals to control weeds," he said. Many strawberry farmers switched their crops to more easily cultivated currants, or they plowed their fields over for hay.

For many years, Bill baled hay on the fields that were former strawberry farms. But over the decades, the climate has changed and the rainy season has increased. "The summers have gotten later and later, and now we can't do anything until the middle of July—we should be out there the first of June. Now it's all turned to straw by the time we get to it, and it's lost most of its protein," Bill remarks.

In the 1960s and 1970s the labor laws changed again, and kids who were twelve to fourteen years old were not allowed to work the farms. Farmers changed over to crops that could be machine-harvested. The last three berry growers were Augie Takatsuka (see Chapter 5), Tok Otsuka, and Yoneichi Matsuda (see Chapter 6).

The most fulfilling part of his years of tractor work is the people, Bill says. "Well, we had a lot of real nice customers. I did tractor work for Fred Ernisee for well over forty years. He kept track of all the acres on the island. I think he told me there were 22,000 acres of fruit on the island. I had a regular circuit in the spring. You get to know a lot of people like family. A lot of them, I've seen their kids grow up and their kids' kids grow up."

Farming is hard work and takes a lot of stamina, Bill notes. "It's like going to work at the factory. You've got to work every day if it needs it. You don't do it just when you feel like it. In the wintertime, there's not much to do in the fields; too wet. But in the fall, you're going out there to do what you can to make it hold over until next spring."

As the agricultural acreage on Vashon diminishes, Harold does the brush cutting and rototilling, and Bill does the land clearing for housing projects. "Now our farm machinery is just sitting out there doing nothing because there's no need for it. So now we're into backhoes, loaders, excavators, dump trucks, and that sort of stuff, because that's all we're doing. I'm going to gravel a driveway this afternoon. The farming is just totally gone," Bill says.

When asked about his future with land cultivation, Bill laughs. "I'm up at that age to retire. I had two or three customers who said, 'Boy, I sure hope you don't retire because we'll be out of business.' Well, they passed away," he says sadly, "and I'm still here."

The Mann Brothers, 11805 SW Cemetery Road, (206) 463-2186

Chapter 2

Eugene and Peggy Sherman: Family Farms

Long-time Vashon-Islanders Eugene and Peggy Sherman recall the many crops that have been grown on Vashon. Photograph courtesy of Eugene and Peggy Sherman.

Longtime Vashon–Maury Islanders Eugene and Peggy Sherman have seen many agricultural cycles come and go in the last eighty years. Their home on Quartermaster Drive is part of 160 acres that had originally been homesteaded by Gene's great-grandfather in the late 1870s.

Gene, born in 1919, grew up on his family's dairy farm (known locally as the Sherman house) in Paradise Valley. He helped his father, Francis Sherman, milk six cows twice a day, seven days a week, for many years. His father had a separator, so he transported the cream to a Carnation condenser in Kent, giving the skim milk to the calves and pigs.

In later years, Francis sold milk to Ed Lande, who picked up milk every day on his rural Vashon delivery route. Don Vermuelen of Tahlequah and Louis Houghton in south Paradise Valley also delivered milk to Vashon homes, Gene recalls.

In about 1933 his family quit the dairy business, then turned to chickens, as did many farmers, seeing that it was more profitable. Gene remembers "lots of chicken ranches, some with 1,000 chickens" around that time. Vashon chicken farmers sold their eggs to England and Peterson, a processing plant located at the present location of the Vashon Pharmacy. There the eggs were graded, packaged, and shipped to market in Seattle.

Gene, with his methodical and enthusiastic attention to detail, rattles off the names of all the other chicken farmers he remembers: "There was Lewis Beall [see Chapter 4], Frank Seigrist on the west side, Elicia Morgan, Will Smith, Andy Jenson on Maury Island, Ed Mooney, Pascov, Nishiori, Seymour Hearst, John Ganley, Cisco, John Brink, Fagan, Clarence Rose, Frank Kneebone, Shane, Tom Dunn, Nakamichi, Tanaka, Floyd Mace, Dumbleton, Wallace Brunson, Reg Cuttler, Bill Clough, A. J.

Lewis, George Jenn, Sam Sakamoto —they're the ones I can think of off the top of my head," Gene says.

Peggy moved to Vashon in 1937, and both she and Gene graduated from high school in 1938. They married in 1941. Their first house was a summer house that rented for $27 a month in 1942. Today such houses rent for $1,200 month. Peggy remarks that in the 1950s and '60s she graded and packed eggs for Ed Mooney on Vashon.

Along with its poultry and dairy enterprises, Vashon has also seen a variety of exotic-animal ventures, Gene recalls, such as chinchilla, ostrich, nutria, mink, and goldfish farms. The goldfish-rearing enterprise was on the north end of the island.

Plants and trees grown on the island have included croft (Easter) lilies, grapes, pears, apples, cherries, peaches, currants, prunes, figs, and filberts. On the corner of Bank and Beall Roads was a prune dryer, across from the present-day location of the water treatment plant, Gene remembers.

One of Vashon's more famous stories is that of Pete Erickson, who developed the Olympic berry, breeding a blackcap raspberry with a loganberry. It was known for its large size and unique flavor, and was a feature for several years at the cafeteria at the Bon Marche store in Seattle, Peggy recalled.

George Sheffield grew a couple of acres of dahlias and sold tubers from his location south of the Mitchell Tree Farm toward Tahlequah in about the 1930s, Gene says. In 1925, Fred Vye grew ginseng in Burton, on 1/2 acre shaded with a lattice-type material suspended 8 or 9 feet in the air, Gene recalls.

Gene remembers that some of the Japanese families grew a large radish tuber that grew 4 inches in diameter and 2 feet long, and was used in a relish-type dish. The Miyoshi family grew 1/2 acre of asparagus and 1/2 acre of rhubarb in the south part of Paradise Valley, he recalls.

All the farmers grew their own hay, and Gene's father grew oats for their horses. But the hay here lacks nutrients, because the weather is too wet for harvest during the time of its high nutrient density. Gene jokes about a western Washington farm he heard about: "The ground was so poor that the cow had to graze forty miles an hour just to make a living."

He remembers two large pear orchards, near the K-2 ski manufacturing site. Over the years, many crops peaked, then subsided. "Once somebody made money on [a crop], somebody else would go for it [too]. It was pears, in the '20s," Gene says. "The market fell off. Things change. Then it was sour cherries—used to be acres and acres of pie cherries. Then it went to peaches, acres of peaches. Then loganberries went through a cycle. Currants. Gooseberries. Before that, strawberries. Raspberries were popular for a while. We've gone through a lot of these cycles," Gene notes.

The labor force changed dramatically over the years, Gene says. The early years saw Native Americans who made their way down from British Columbia through the fields at La Conner, then south to Vashon to follow the crop harvests. "They loved it—the whole family would come down and pick berries," Gene recalls. "Then our government, in all its vast wisdom, decided that every camp should have hot and cold running water and showers and flush toilets. Well, the farmers couldn't [afford to] provide that. The last ones to have Indians down here were the Larsons—Kenny Larson on Maury Island. He had about 160 acres of currants. And they were here up until the middle '60s again, every year. Many farmers, being forced to put in these sanitation requirements, quit farming. [The expense of the upgrades] put them out of business."

Some of the Vashon youth harvested crops too; up until the late '40s or early '50s, school let out in the middle part of May, Gene recalls. Such local day labor didn't require the expense of overnight lodging facilities. In those days, kids earned their luxuries. Says Gene, "Our kids picked bicycles."

Today, Gene and Peggy tend a small garden plot in the yard of their home on SW Quartermaster Drive. They grow potatoes, corn, tomatoes, sunflowers, green beans, figs, berries, and filberts. Gene is most proud of his finger potatoes, a long, narrow, small potato 1 inch by 8 inches long that he has developed by selective planting for several years.

Gene's view of the future of farming on this little island is grounded in his historical perspective from nearly eight decades of agriculture on Vashon:

"Well, the kids of the farmers don't want anything to do with it. It's too much work. So it's being lost. And there's not enough income for the laborers. They can make more money at Boeing, for instance," Gene observes. "I don't think there's any future in farming except for a few small cottage-industry-type gardens," Gene says.

"And farmers who work somewhere else to support the farm," Peggy adds.

"Taxes are so high now," continues Gene, "and equipment costs so much money, too, and the fuel to operate it."

When asked whether he favors the idea of community-supported agriculture, which has seen growth on the island in recent years, Gene replies: "I think it's great. There's so much commercially grown stuff that is so infused with different chemicals and hybridized so that they don't get various things wrong with them. To grow [crops] naturally, I really go for it."

Chapter 3

Arthur Tom Lorentzen: Norwegian Settlement at Cove and Colvos

(Arthur) Tom Lorentzen remembers his youth "smashing ripe peaches into his face" while working hard on Vashon's fruit farms. He recalls the small Norwegian farming and fishing settlement near Cove and Colvos. These days are easier, gardening as a hobby, and spending time with his horse, Stopper. Photograph © Kathleen Webster.

Johnny Lorentzen and Leo Larsen hanging Olympic berries on August Johansen's farm, about 1930. Photo courtesy (Arthur) Tom Lorentzen.

(Arthur) Tom Lorentzen's neighbors to the east, Maren and Hans Hamer relaxing in their home around 1920. Mrs. Hamer was the local midwife. Photo courtesy (Arthur) Tom Lorentzen.

Born in 1928, Tom Lorentzen, teacher, storyteller, farmworker, recalls the Norwegian settlement at Cove and Colvos. Back then, in the '30s and '40s, the men worked hard as fishermen nine months out of the year while their wives toiled the fields, managing the home front and the children.

Tom gives a memory tour of the Cove and Colvos settlement, an area within a 1-mile radius of where he lives, from his home to Colvos dock, 1 mile to Cove, less than a mile over to Masa Mukai's field and processing plant, and 1/2 mile to the water. Leaping around, gesturing to his collection of historical photos, his painstakingly detailed family tree, and other memorabilia, Tom rattles off facts, tall tales, and jokes—a master storyteller.

"The basic farm was 10 acres facing on 123rd or Cove Road or the Westside Highway," he says. "They all had a barn, a cow, many had a horse, and usually, like my dad, Leonard John Lorentzen, they raised one or two pigs a year, and a bunch of chickens. For the families that had fishermen in them—my dad and my brothers were fishermen—farming was supplemental. You always had something—the potatoes, the chickens, the pig to kill each year or a calf. Everything was salted down; we didn't have refrigeration. Those who didn't have a fisherman in the family had to make their living on two things: either chickens or berries."

From 1980 to 2001, Tom lived with his wife, Bea, in the old Krockset place, 10 acres of mostly hillside that takes five hours to mow, with a complex drainage system that still works ninety-five years later. In 2001, they remodeled the Krockset place for their son Rolf, daughter-in-law Lisa, and grandson Andreas, and moved back to the old Lorentzen place next door. The former owners of the property, Johanna and Ivor Krockset, ran 3,000 pullets a year, and had a field of loganberries down below, Tom recalls. Overlooking both properties is a neat, sturdy barn that still has the rope and pulley hoists for the hay cut from grassy pastures that used to be stored as loose hay in the loft.

He then narrates a vivid memory tour of the neighborhood of "probably the most intensively farmed area of Vashon," consisting of small homesteads, starting north from the Lorentzen place. "The Molviks next door had chickens and berries; they did a lot of wood cutting, and their sons were also fishermen. Next door, George Olsen, carpenter, farmer, ran chickens. Then the Lelands, who had lots of chickens and a few berries. Next place north, August Johansen ran tremendous numbers of chickens and had lower fields of Olympic berries; earlier it was loganberries but then he shifted to Olympic berries. Next door, Ellingsens ran chickens. Thomsens the same. John Olsen was the last one on that end of the section, and then the Walls family was beyond SW 148th as you go north. Next door, Lawrence Walls had berries and a lot of chickens. Below the road on the west side just above the Colvos dock was August Johansen's son, Barton; he ran chickens, but mostly he had berries.

"As you come south along the road, the most successful farmers on the whole west side—the Seigrists—had a great number of chicken houses and ran a lot of chickens. Right down there by that old white Lutheran church was Conrad Christensen; he ran 20 acres of Perfection currants and he ran three big chicken houses; he had strawberries and loganberries. We picked for him every year, my mother and my sister and I. I can even remember being put to sleep on top of a berry flat under the bushes, so my mother could pick," Tom laughs.

"We always felt happy when the loganberry season was over and we could pick currants, because we could make so much more money on that—they were hanging like grapes, and you could just pull them off and fill up a flat," Tom remembers.

"Next door to Conrad Christensen and his wife, Margareta, was Mrs. Reine. Her husband had been a carpenter, and he had died in 1930. So she was a single woman farming 10 acres, if you can believe it. She had about 4 acres of pie cherries at the road and then 4 acres of loganberries; below that she had currants, chickens, a cow, and a horse—we borrowed the horse all the time. Mrs. Reine was the most successful single woman I know running a farm all by herself. She hired my brothers to help with the berries and clean the chicken houses. Whenever she would go to Vashon, she would pick us up for a grocery trip, in a big fancy Buick I think it was, in the early to mid-'30s.

"Next to her was Mrs. Huseby; her husband, Bernt Huseby, had drowned the same time two other neighbors did, when the *Orient* was hit by a steamer up off the north end of Vancouver Island. Another was Andrew Lokke across the road. At least two others were survivors, Mr. Andrew Ellingsen up on the hill and Ed Langsness, a relative of Mrs. Reine from Norway; I forget how many went down with the *Orient*. Mrs. Huseby was left a widow with four kids—two daughters and two sons; one daughter married my oldest brother Ingvald (we called him Inky), and she still lives down below here: Ragna Huseby Lorentzen. Her mother managed that place and they ran chickens, they had loganberries and fruit trees, and of course a cow and frequently a pig as well.

"This way from the Husebys, the first place to the north of the [Colvos] store, Mr. Rockness probably had as many chickens as anyone here on the west side. Rockness, another Norwegian, had no other income [besides chickens]. But, boy, he did a massive job. I used to stop by as a little kid and look in there at all the feeding and cleaning, and everything just absolutely perfect."

The general store, today known as the Colvos store, passed through the ownership and management of several families—Televiks, Trones, Sextons, Kresses—Tom recalls. "Behind [the store], the Trones farmed 5 acres, everything from cucumbers to berries. Right behind them was Al Roen, who grew 5 acres of the best peaches around. Coming up this side of road is what is now the Whitman-Beardsley place; it was the Dannevig place—he was the one non-Norwegian. He ran 6 or 8 acres of loganberries and chicken houses galore—probably ran 5,000 or 6,000 pullets a year.

"Up from them, directly across the road from us, was Andrew Lokke. They had a big barn right by the road and three big chicken houses; they ran 5 acres of loganberries right here on Cove Road, and fruit trees and a vegetable garden. They cut an awful lot of hay because he had a horse and I think he had two cows. He actually had a sidebar mower that he pulled behind the horse to cut his hay, and so he would cut other people's hay for them, too. But like most people, my dad cut his with a scythe; that whole area from the swamp on down to the road, each year he would just cut it—6 inches, a step at a time. I've still got the scythe. My weed eater isn't as efficient, I'll tell you that.

"If you keep going south, around the corner to the southwest, the Wangs had chickens and berries. Across the road from them was Hansens—berries and chickens. Next to the Hansen place they had the fanciest chicken houses around, and now they're owned by Warren Beardsley—tremendous chicken houses built just beautifully. Next to him was a friend of my dad's from Norway, Aasman Johnsen; he had three long chicken houses.

"Then you get into fisherman territory," Tom says, showing a picture of the Cove dock, "this dock right here, the old Cove dock. Up on the hill above the dock was Conrad Andersen and all the Edwardses. The Edwards family owned much of the land around, and they owned most of the boats down there. Jacob Edwards was a sponsor for a lot of the Norwegians—he signed for my dad to come out as a halibut fisherman for him. A lot of these people came from Norway on his recommendation and his sponsorship.

"The boats tied up there and went to Alaska from here. My dad and brothers all fished. Edwards and his boys owned all sorts of boats, but that all came to an end in 1943. Two of their boys drowned out there in a big storm in '43. Four of our neighbors' sons drowned, and two of them were Edwards.

"South from the dock up on the hill above the Cove dock, there were also many chicken farmers. Bruner had an enormous number of chickens. Next to him was Albin Sundberg, our Sunday school superintendent; he had huge chicken houses of two and three levels. East of our place if you go up over the hill now along the upper road past Kvisvik's, you come to the Zarth farm of 40 or 60 acres, one of the biggest farms, with horses, berries, chickens, and lots of cows as well.

"Then as you come south along 115th, you come to the biggest strawberry patch that we picked in as kids. This was Masa Mukai's strawberry field, which is just due east of us here, up on 115th. It was all cleared by Tjomsland and Masa Mukai. Right where the airfield is now, it was 60 acres of strawberries, as far as you could see, all the way down to the Steen mill at the beginning of Shinglemill Creek all the way up to 115th. Now it belongs to the Catholic archdiocese, St. John Vianney Church property, and the airport [see Chapter 7, Masahiro Mukai's Vashon Island Packing Company].

"We picked berries for several neighbors, and always we picked for Mrs. Reine; she hired my brothers and they hung the loganberries. You have to cut down the old vines and hang the new ones up. We hung vines for Johansens, Christensens, Lokkes—anybody who needed someone to hang berries; my brother Johnny was an expert hanger," Tom laughs and shows a picture of himself and a buddy hanging berries in the field, then shows a picture of the local midwife, spinning wool, a contrasting view of life inside the homes in those days.

In the early 1940s, during his youth Tom worked on some of the fruit farms outside his neighborhood, following the harvest season. "It was strawberries first, the end of May, then we'd work to loganberries, and I guess the currants would have been the last, and so I picked them a lot of different places. When I was in high school, Betsy Wax [of the Wax Orchards family; see Chapter 8] hired some of us kids. The Waxes had about 40 acres of peaches; they had gooseberries, also currants, and about 80 acres of pie cherries. They hired my sister and my mother to work in the peach packing plant. I worked for them collecting the peaches in the orchard, driving a tractor and picking up all the buckets they would set down by the trees. I would load it on a sled and take it into the shed, and I'd load it on the deck. Then I would go with Mr. Wax to town each morning and he would sell the whole darn truckload, to Pacific Fruit and Produce in Seattle or whoever gave him the best price.

"Once they didn't give him his price, and we drove all the way to Puyallup to sell the peaches there. My mother and sisters, all the ladies who were packing, would work there until midnight and then come back the next morning and start again. My mother and sister also worked for Masa Mukai in the Vashon Island Packing [Company; VIPCO] plant, working on the belt, picking out the bad strawberries. During [World War II] I also picked for Shattuck, who was managing Mukai's and Sakahara's fields right up there east and south of the VIPCO plant.

"I worked for the Waxes after school cultivating gooseberries, driving a little Cletrac [an abbreviation for Cleveland Tractor Company] down between the rows—and tearing up all the gooseberries when I would miss a turn," he laughs. "I liked the Waxes, I liked working down there. It was fun. But it wouldn't be all healthy. We would spray the cherries with lead arsenate. I would sit on top of Betsy's 500-gallon sprayer and she'd run the water in, then I would dump several packets—5-pound packets—of this gray lead arsenate into the sprayer. I would be stripped to the waist, and as she would drive through the orchard, I would sit in back spraying up and down through the cherry orchard. I was absolutely soaked [indicating his chest and belly] in lead arsenate, all day; I would spray her, too, because [he laughs] I wasn't always very accurate. That was the standard spray, about 1943," Tom recalls.

Tom went on to work in construction, then the shipyards, then, unlike his siblings, went on to college. College was not without sacrifice. To afford it, he shared a 12-foot diesel-leaking trailer with a friend. His dad, back at home, took care of Tom's beloved horse, Lucky. But since no one rode him, they decided to sell him. Tom pauses and rubs his eyes at the memory. After he graduated from Washington State University, he taught English composition and other subjects in Seattle schools for seventeen years, and then taught at South Seattle Community College for another twenty-two years.

Today, if you stop over at his house, you'll see his two small, neatly managed gardens, what he calls "God's little acre, where the deer and the antelope play."

"So I'm just a gardener now, and I prune about seventy-five fruit trees, apples, pears, plums, every blinkin' kind of tree you'd want.

Lots of nut trees. We don't use them; they go to the ground—the squirrels love 'em, and the jays—Chinese chestnuts, oak, basswood, or American linden down by the road there, and lots of different fruit trees. Takes me a month to prune. And climbin' around in branches when you're seventy-two years old, looking down, wondering, 'why am I up here?' It's not very much fun, looking down, waggly knees. ...I'm not sure I'd recommend farming to anyone.

"Still, I told Augie Takatsuka [see Chapter 5]—I stop by and visit him every day—'Augie, I just drove up here, through the whole [community of] Colvos, up along through the upper Colvos area, past the Zarth's, past Mukai's, all the way to your place, and I saw not one person working in a garden.' Nor did I see one plant raised for food except in my own driveway.

"In the old days when daylight came, Mr. Lokke was out with his horse, Molvik was cuttin' wood, Leland was cleaning chicken houses, and today I don't see one person working on anything outside. Once in a while, you might see someone trim an old bloom off a rhododendron. They're into gardens that are for beauty. Well, beauty is in the eye of the beholder. I like straight rows.

"Doc Eastly drives by with his big [draft] horses and he always has a load of people with him. He'll stop and say [in a loud, gravelly voice], 'Now look at that unimaginative garden there, look at that, the rows are all straight, there's no damn weeds in it. What's wrong with this guy?' just joking away. But when I invite him in to the pea patch, he goes absolutely bonkers if he can pick some peas on his own.

"But yeah, I do it because I like to do it, as long as I'm still strong enough to do it."

Chapter 4

Tom Beall Jr.: The Beall Greenhouse Company

Tom Beall Jr. recalls the floral abundance of his three-generation family rose-and orchid-growing business. Blackberries have overgrown what's left of the frames of the greenhouses, shown in the background. Photograph © Pamela J. Woodroffe.

An aerial view of the expanse of the Beall Greenhouse enterprises with Maury Island (part of Vashon-Maury Island) in the background. Photo courtesy of Vashon-Maury Island Heritage Association.

Wallace Beall tends flowers in the Beall greenhouse. Photo courtesy of Vashon-Maury Island Heritage Association.

The Beall's Pedigreed White Leghorn poultry farm. Photo courtesy of Vashon-Maury Island Heritage Association.

As a young man, Tom Beall Jr., born in 1946, grew up on Vashon with the memories of the overwhelmingly beautiful, sweet scent of roses from his family's business, Beall Greenhouse Company, established on 28 acres in 1888 (see "History of the Beall Greenhouse Company" at the end of this chapter). From age five, Tom Jr. worked alongside his father, his five brothers and sisters, and longtime employees, learning all levels of the business.

Today, in his home across the street from what remains of the wood-frame greenhouses, Tom Jr., soft-spoken and humble, recalls, "I would run across the dirt road here every day and hang out around the greenhouses because I enjoyed being with all the people over there, and I could be with my dad all day long. It's a great way to grow up. I started pulling weeds at an early age," he says. "That's where you learn the business from the ground up," he smiles.

Tom Beall Jr. shares the memories given to him by his father, Tom Sr., of the early days when the greenhouses mainly grew vegetables and he took the crop by horse and buggy to the dock, where it was taken by boat to the mainland. "My father told the story of being able to ride in the wagon pulled by horses down to Vashon Landing with the produce, putting it on the boat, and coming back up the hill. They brought back supplies for the cookhouse and lumber for the greenhouses. The horse would always make the people walk back up the hill. The horse was very smart. He didn't mind going downhill with the flowers and fruit and produce. But coming back up, everybody had to walk. Today, that trail now is just a very small trail down through the woods. The road has now been abandoned for years," Tom says. Today, Vashon Landing is east of the Thriftway shopping center, at the end of Soper Road, where the road meets the water.

Tom Jr. remembers his father's tales of cutting cordwood as a youngster and firing the boilers to heat the greenhouses. More greenhouses were added over the years to keep up with the market demand for fresh-cut orchids, and then they expanded into growing roses. Additional greenhouses were built, and it became more practical to heat them with coal through the 1930s. After World War II, they bought huge ship's boilers that burned "black oil," a thick, heavy fuel.

In the 1950s, when Tom was five, in an effort to diversify its crops, the family bought truckloads of Christmas trees from Port Angeles tree farms and flocked thousands of trees for sale through local florist shops.

Tom recalls with fondness the early 1960s, how much he enjoyed working with his brothers and sisters and his father, and the other twenty or thirty people on coffee break who were part of his extended family, all of whom were grateful they didn't have to commute off-island to make a living. For many years, the Beall Greenhouse Company was Vashon's largest employer. "You'd never know it now. It's history that's almost forgotten. We put many, many kids through college," Tom remembers, adding that they also hired Vashon teachers to work in the summertime.

When he was twelve or thirteen years old, he was tall enough to cut roses three times a day. After, he learned to use pipe wrenches and work on the boilers. He drove tractors at age fifteen or sixteen, putting in water lines, steam lines, doing the electrical work. The operation had huge boilers and their own electrical generating plants for standby electricity, Tom says. On an island, you had to be self-sufficient in case of winter power outages, know how to fix everything and not wait for someone from the mainland. "You could lose the entire crop in eight hours of freezing weather, without the boilers to heat them," Tom says.

In high school, when he was old enough to drive, he worked weekends and trucked flowers to Seattle-Tacoma International Airport. "I learned a lot about hard work ethics, putting in a good twelve-hour day, seeing the fruit of all your hard work, bringing in the crop on time. Seeing the customers' faces light up when they get their flowers. It was a great place to grow up," Tom recalls.

Tom Jr. was eager to learn the business. His father recognized his potential and set him to apprentice with several workers for a month or two at a time. "That's what was neat about my dad. It was very rewarding to learn—how to cut flowers, how to put together steam pipes, build a greenhouse. By golly, those guys knew what they were talking about. You learn a lot if you just take the time to listen," Tom says.

After he graduated from high school, Tom Jr. attended Yakima Valley College, then married Nancy Evans in 1964. They have three children, David, Chris, and Jackie.

The family continued to build more greenhouses and renovate the old ones, up to the mid-1970s. At its peak, Beall Greenhouse Company had ninety-two greenhouses—a half million square feet of wood-and glass buildings, Tom Jr. recalls. At its peak, there were 112 employees in the summertime and 60 employees in winter. This compares with today's major manufacturing employers on Vashon Island, such as Pacific Research's 65 employees and, until recently, K-2's 500 employees.

"In the 1970s, my father and his brothers didn't realize that the fuel crunch would last so long. No one did. And you had to keep your customers and develop a customer base that would buy your flowers year-round," Tom says. The production costs, especially heating and labor, were increasing. "We would make very good money in the summertime, and then we would spend it in the wintertime, because we would have to pay the fuel bill. Each year we'd make it a little bit ahead, then we'd lose ground."

In 1976, Tom Jr. returned from four years in Bogota, Colombia, where he had built the largest rose exporting business in the region, to Vashon to assist his father. The company split into two corporations, with Tom Sr.'s brother John managing the Colombia plant and Tom Sr. managing the Vashon greenhouses.

Tom Sr. retired in the early 1980s, and Tom Jr. carried the business on, but his tenure as owner and manager was to be short-lived. Already Canada, with its large natural gas fields to fuel its greenhouses, could produce roses of similar quality for half the price. The nightly fuel bill for the Beall Greenhouse Company was $600. It became more and more difficult to make that cost up in sales. And winters were getting colder.

The family researched every way to keep the business going. "We investigated every crop and every avenue of what we could make work in the greenhouses. We brought in experts," Tom recalls. The last three years of the operation, they were forced to lay off forty-five people and shut the greenhouses down one by one. At the end, they were down to four employees and ten greenhouses.

"It's very difficult to let people go after they've been working for you for so long," Tom says with quiet regret, his eyes misting. "Especially when you've had [the company] three generations, and you have to be the one who says, 'Well. There's no alternative. I've looked at everything.' But we had to say, 'You know we're sorry, but we just can't keep the operation going.' Many other greenhouse operations in the area had closed their doors, having experienced similar challenges."

Tom Jr. slowly dismantled the business. The Tacoma warehouse was sold in 1980; the orchid department was sold in 1983. In 1986 Tom took a job with Pacific Research, locally known as the Bone Factory, which manufactures plastic bone models for doctors to practice surgery. In 1989, 101 years after its founding, the 28-acre site was sold.

"It's been about ten years since I've sold it. And every day I still think of the good times we had over there, all the crops we brought in and all the money we paid to the people who lived here on the island. We just had to say, 'I gave it my best shot.'"

Asked what he would do differently if he had it to do over, Tom thinks it over for a long time. "I would close sooner, and not try and struggle for the last five years, and recognize the existing market competition and the fuel challenges," he says.

"But I'd also say the mom-and-pop family-run businesses are a really important part of the American tradition. Ninety percent of the people in today's economy don't understand what a farmer really goes through twenty-four hours a day. We spent twenty-four hours a day in the greenhouses during storms. It starts at 4:00 in the morning and ends at 6:00 at night. There's nothing like walking into a cooler of fresh-cut roses, smelling the fragrance, seeing thousands of flowers coming on bloom, and starting to cut. And five hours later, you've taken out 3,000 or 4,000 blooms for Valentines' Day. It's very rewarding. But it's very discouraging when it would come time to balance the books at the end of the night and you've lost $10,000," Tom says.

Tom's advice for aspiring growers? "New farmers need to know that their time is valuable. There's going to be a start-up period. They've got to be able to pay themselves enough to get by on. They've got to have their own niche, their own marketing setup; they've got to make their books balance, and not just do it for the love in their heart.

"Still, I'm glad that I had the experience. It was really very rewarding to me, and to my whole family. I've seen them carry on with their children and the work ethics. You work hard and most years you are rewarded for it."

History of the Beall Greenhouse Company

Compiled by Mary J. Matthews

Hilen Harrington, who hailed from a greenhouse business in Norwich, New York, in 1878, was the original founder of the H. Harrington Company Greenhouses on Vashon. He moved into the log house on the site, then returned to New York to fetch his wife, Ella, and ten-year-old daughter Jessie.

By 1888 Harrington Company boasted nine greenhouses, erected on a plot of land still scattered with the trunks of the old-growth forest. The young business was established on firm financial ground with the advent of the Alaska gold rush, when adventurers and purveyors alike paid handsomely for H. Harrington Company cucumbers and tomatoes. Harrington's success was advertised for all to see in 1892, when the family moved from their humble log house to an elaborate Queen Anne–style mansion built right at the cabin's front door.

As early as 1896 Harrington had branched out from vegetable growing to floriculture. In 1898 Harrington's business received a financial boost when capitalist Lewis

Cass Beall Sr. moved to Vashon Island. L. C. Beall Sr. was a prosperous merchant from Beltsville, Maryland, who had attended the Maryland Agricultural College and established himself as an authority on horticulture. In addition to this experience, he brought some capital investment as well. His wife, Sarah Jane Corner Beall, was a southern belle from Georgia, and her elaborate manners and strange accent soon earned her the sobriquet of "Madame Jennie" from Vashon Island's rugged frontier populace. The Bealls and their children—Lewis Cass Jr., Connie, Wallace, Magruder, and Allen—first settled in a house near Center for about three years. In 1899 the Harrington and Beall dynasties were intertwined when L. C. Beall Jr. and Jessie Harrington were married.

Following this marriage, L. C. Beall Sr. and Hilen Harrington formed a new partnership beneficial to the company. On March 3, 1902, the H. Harrington Company was incorporated with capital stock of $18,000, with H. Harrington, L. C. Beall Sr., L. C. Beall Jr., and Wallace Beall as trustees. That year, L. C. Sr. and Jennie built a two-and-a-half-story house east of the greenhouse complex.

By 1900 the business had grown to fifteen greenhouses, a barn, a cookhouse, workers' housing, and other buildings. The years following the turn of the century saw the business expand; the H. Harrington Company operated two retail stores in Seattle. In 1912, at age sixty-one, Hilen Harrington retired from the greenhouse company and sold his interest in the plant to L. C. Beall Sr.

and his sons Wallace and Allen. The business was renamed the Beall Greenhouse Company.

Upon Harrington's retirement, L. C. Beall Jr. also left the greenhouse company and embarked upon a new enterprise: the establishment of a pedigreed poultry farm. Consistent with the Beall family's other commercial successes, the productive hens of the Beall Pedigreed White Leghorn Farm soon became world famous—even going so far as to win first prize in a national egg-laying contest in 1921.

By 1915 the Beall Greenhouse Company had thirty-six greenhouses covering an area of more than 110,000 square feet and was the largest flora plant in the Northwest. At the time, it employed ten people and consumed 1,400 tons of coal annually. The years around World War I culminated a half century that saw the demise of the founders of the company: Hilen Harrington died in 1916, his wife, Ella, died in 1922, and in 1926 L. C. Beall Sr. died. L. C. Jr. took charge as president of the Beall Greenhouse Company, with Wallace serving as its director. During the decades following World War I, the company produced many prize-winning orchids, roses, carnations, camellias, and gardenias. In 1942, Jennie Harrington Beall, the last of the company's founding generation, died.

In 1946, an 8-acre plant was established in Palo Alto, California, devoted to roses and managed by one of Wallace's three sons, John. Another of his sons, Tom Beall Sr., became general manager of the Vashon plant, and Wallace's brother Allen ran a retail florist shop in Seattle. During World War II, the company became well known for its role in saving rare English orchid stock from the Nazi threat; Wallace purchased more than 8,000 orchids and had them shipped on six different boats to ensure that at least one would make it safely. Even with these precautions, only 5,000 plants made it alive.

After World War II the company focused on the production of orchids and roses and continued to expand under the leadership of Wallace and his sons. Wallace's third son, Ferguson, known as "Fergie," returned from serving as a fighter pilot to become the Beall Greenhouse Company's orchid expert. Fergie experimented extensively in orchid propagation and hybridization, cross-pollinating to create different hybrids. Orchids need a nighttime temperature of 68 degrees Fahrenheit, while roses needed to be kept at 62 degrees at night. So the two species were housed in separate greenhouses, with their own lighting systems. Most of the flowers were shipped to the Midwest, where they were in high demand due to the Midwest's overly hot summers and too-cold winters. Palo Alto roses were shipped all over the southern half of the United States.

Fergie served as president of the American Orchid Society in 1959. By this time, the Beall Greenhouse Company was producing 500,000 orchids a year, and they were featured on local television in "Success Story." During the 1960s, they were one of the largest producer of roses in the western half of the United States.

Wallace's death in 1964 marked the end of an era for the Vashon plant of the Beall Greenhouse Company. With increasing production costs, especially heating and labor, the company's founding location on Vashon Island could not compete, and the Bealls began to consolidate and look for new locations to grow roses. In 1972 Tom Beall Jr. went to Bogota, Colombia, to build the largest export rose-growing operation in Colombia, and that same year the company closed its Palo Alto plant, which had become surrounded by housing developments. In 1976 Tom Jr. returned to Vashon to assist his father, and the company split into two different corporations, with John managing the Colombia plant and Tom Sr. managing the original Vashon location. The plant in Bogata was later closed.

Despite efforts to make the Vashon plant economically viable, the 1980s saw production costs continue to rise, and Tom Jr. slowly dismantled the business. In 1989, 101 years after its founding, the long and prestigious history of the Harrington Beall Greenhouse Company came to an end when the Vashon Island plant was sold to its present owners, Charles and Nancy Hooper. Currently nothing is growing on the property.

Part II
Strawberry Fields Forever

Vashon's Strawberry Industry and Japanese-American Farmers

Chapter 5

August "Augie" Takatsuka: From Strawberries to Christmas Trees

Augie Takatsuka, one of Vashon Island's premier strawberry growers, survived weevils and war, and recalls the hard work and high risk of farming. Photo © by Kathleen Webster.

A typical hot summer day of Vashon's past, when its landscape was dotted with workers harvesting fruit. Photo courtesy of Vashon-Maury Island Heritage Association.

In 1911 Augie Takatsuka's father, Yohio (George), and his mother, Aya Taki, arrived on Vashon and made their home in the Gorsuch area. Augie, born in Portage in 1921, was one of six children. His parents farmed 10 to 15 acres, 5 acres at a time, with one horse. They packed berries into boxes that bore a special stamp and loaded them onto the Mosquito Fleet boats headed for Seattle, where the cargo would be transferred to a freight company that took them on to a commission house such as Pioneer or Ryan. The Mosquito Fleet boats were steam-engine vessels that carried passengers, freight, and mail to and from Seattle, Tacoma, and Colvos Passage before highways replaced them as a means of transportation.

Augie is a quiet, matter-of-fact gentleman who lives in a modest home on the remaining 27 acres of his farm on SW Bank Road. He recalls that his family grew strawberries for several years, until 1919 when the strawberry weevil plagued the crops. "The weevils lay eggs around the plant that hatch out the following spring, then they eat all the roots off the plant," working their unseen damage underground, Augie explains. "Sometimes there's loss when they eat the old root—they just kill the plant. You'd have a huge plant like this [motions with his hands to illustrate] that's been going all fall, but the root system is being eaten up. In the summer, on the first hot day the plants just wilt."

By the time Augie was old enough to help his parents farm, chemicals had been discovered that would combat the weevil. "You've got to bait the weevils," he says. "When I started farming, there was a bait made from apple pomace soaked in silicate something—I forget the name of the poison—that you applied over the plants at a certain time. It had to be carefully timed. The beetle is in a soft stage for about three weeks, and when it matures, then it starts laying eggs. So you have just a three-week period when they're feeding to bait them before they lay their eggs. You had to keep watching all the time. You knew that you could work all year, but if you didn't get this bait on in time, why, it was all for naught. Everything wilted and died."

In the late 1930s and early 1940s, Native American families were hired to help with the harvest season, Augie remembers. "The Indian families used to come down as kind of their summer migration route. A lot of them came and pitched a tent on our place and picked berries. Some of them didn't even have tents, and they'd just say, 'Well, we'll just go camp in the woods.' It was very primitive. The health department would just throw up their hands in despair. The Indians would say, 'Well, if you chase us off, we'll just go off in the woods and camp there.' And they did," Augie recalls.

After Augie had graduated from high school in 1938, he leased a piece of land near his family's home in 1942, and planted strawberries there. (In those years, laws prohibited Japanese people from owning land, so they farmed land they had leased from other owners. See Chapters 6 and 7.) This was to be Augie's first crop on his own. Rumors were spreading that strawberries would bring five cents a pound that season—a high profit in those days, Augie says.

It was World War II—and announcements of the evacuation of Japanese-American people were posted on Vashon's telephone poles.

"In school, I'd been taught to salute the flag, just as American as the next guy. [U.S. government officials] said, 'Well, they're going to move you off.' And I said, 'Hey, we're just a couple weeks from harvesting [strawberries].' We had to leave in May. I remember eating a few strawberries right before we went," Augie recalls.

"When the day came, everybody made their way to the Vashon dock. All we could carry was one little suitcase each. If we had pets, we had to leave them with a neighbor. Most of them got shot, I guess, after we left. That was it."

His family boarded the ferry to Colman Dock in downtown Seattle, where bayonet-armed guards were waiting. The guards herded the people onto trains, which took them to a temporary camp in Pinedale, California. Along the way, when the train passed through towns, the guards drew the shades. "They didn't want people to see us going by," Augie recalls.

After two days without food, his family was taken to a barrack they came to call chicken coops. Later that summer, they were transferred to Tule Lake, a permanent camp in northern California. Some of the Japanese families from Vashon were in the same block with his family, Augie says, while others were scattered around the camp.

It could have been worse, Augie remarks.

"I used to think, had there been another person in command of the camp, we could have just disappeared, [been taken] into the desert...and it gives me the shivers—it could have been like the Holocaust. Everyone said, 'We have to do this for the war effort. We must do whatever they say, give them the least amount of trouble and work with them.'"

Rather than face a future at the internment camp, Augie volunteered for the 442nd unit of the Nisei Regimental Combat Team. (Nisei is an American-born and-reared child of Japanese immigrants.) He trained in Camp Shelby, Mississippi. During his army career, he took care of the horse stables, including General Eisenhower's horse. Augie saw combat duty in Italy, and got a severe case of trench foot. He was hospitalized for two one-year stints, at Madigan Army Medical Hospital in Tacoma, and at the

Veterans Administration Hospital in Seattle on Beacon Hill, before losing part of one foot.

"You hear these stories [that] they just whack [your feet] off. Well, they don't. You have to beg them to do it, the pain gets so bad," Augie remembers. He couldn't stand to have even the bed sheets touch his feet, he recalls.

After World War II ended in 1945, Augie returned to Vashon and leased a piece of land across the street from what is now McMurray Middle School. In 1947 he harvested a bumper crop of strawberries, and soon after bought the land. Back then, good, cleared land could be bought for $350 an acre. The National Fruit Cannery financed Augie's plants, fertilizer, and equipment, and with that and his $100-a-month GI (government issue) stipend for having served in the Army, he made it through the first year.

Augie hired Indians from British Columbia for his labor force, and over time built living quarters for the migrant workers from tent frames, then constructed small cabins. "Things were changing—the health department was getting very pushy," Augie shakes his head, recalling the living conditions he and his family were forced to endure when they were interned during the war.

"To this day, I resent the fact that the health department got all over me, and the same government stuck me in a tarpaper cabin with forty other people when they evacuated us. Well, they said that was a 'necessity.' And I said, 'And now you're telling me if I don't put up a motel-like building, I can't keep these migrants here?'

"I used to have to bite my tongue when they set the rules and regulations down. I didn't want to tell them off—that didn't do any good. I just got myself in a lot of trouble; when I did spout off to them, it cost me many thousands of dollars."

It took careful planning to make a living strawberry farming. In the winter, Augie would predict how many pickers he'd need for the coming season and hire a "picking boss" to travel to British Columbia to recruit Native Americans. "You'd have to be a good judge of how many tons of berries you might get, so you might say, 'We need seventy-five grown-ups to come down.'" He could count on getting four tons of strawberries an acre, and the rule of thumb was that one adult farmworker was needed to harvest each ton. Cash advances for the bosses were common. The bosses often stayed on during the season in supervisory positions.

Though Augie grew mostly strawberries, he added currants to increase his profits and retain his labor force with a more lucrative crop.

"The money isn't [made during] the peak of the fruit; your money is in the tail end of it when [the fruit] is getting thin. That's when you've got all your expenses paid, and if [the farmworkers are] starting to jump ship on you in pursuit of other farmers' currants…if you had a nice field of currants sitting there, well, they stayed on," Augie recalls.

Over the years, Augie sold his berries to Vashon Island Packing Company (VIPCO), on Vashon; Northwest Berry Packers on Bainbridge Island, Valley Packers in Puyallup, and other food canneries in Seattle and Mount Vernon. In the early '60s, he furnished fresh berries to five or six Lucky grocery stores, and Big Bear grocery stores. Then as those stores were absorbed or replaced by larger chains, the larger grocery stores wanted 2,000 flats of berries just one day a week. With crops that needed more frequent, smaller harvests, Augie couldn't meet their demands. "Strawberries don't wait. We had to pick every day," Augie says. Then the chain stores began ordering berries from California, and

Vashon farmers couldn't compete with their prices and quantities. Augie continued to supply the small independent grocers instead.

"We were picking fresh berries—[the farmworkers] would get up at daybreak and pick fresh market flats, then the rest of the day they would pick the canners [berries used for preserves and such]. I'd have to be on that 9 a.m. ferry at the South End [Tahlequah, that connects Vashon to Tacoma]," Augie remembers. "Sometimes I'd make a loop clean around through the Tacoma area, Kent, and up Beacon Hill (in Seattle).

"That was when I had about 60 acres of currants and strawberries, in the '50s and '60s. Those were the peak years," Augie says.

In the mid-1960s, Augie had agreed with the health department to implement an incremental ten-year upgrade of his camp for migrant workers, including showers. But then one season, a King County Health Department inspector required additional improvements, such as rewiring, and Augie couldn't afford it. The health department told him he couldn't farm that year.

In the mid-and late 1960s, most Vashon farmers began to hire Seattle youth to harvest the crops during the day, rather than provide housing for overnight migrant farmworkers. The kids on Vashon weren't plentiful or hard-working enough to support the farms, Augie says. "For a number of years, we hauled kids from Seattle by the hundreds. But we got complaints, because it would get hot and kids would [leave] the fields [early and] walk down the highway [to the dock], some of them getting into mischief, bashing mailboxes and this and that. If you fired them, you had to take them down to the dock, or they'd get into more mischief," Augie says.

"It just got harder and harder to retain workers," Augie says. "I could raise the strawberries, but I couldn't harvest them. I've had this whole field back here never picked for lack of pickers. [The berries] just laid out there and rotted. We did that for a number of years—good strawberries, fruit so heavy. Soon it got to be, well, I might get a third of them picked, and that wasn't enough to keep me in business."

Augie recalls a conversation he had with a young man he hired to pick berries; the youth said, "I'm sorry, but I'm gonna get a job flippin' hamburgers."

"I pay them more than they're making by flipping hamburgers, and they [still] quit. I asked him, 'Why?' And he said, 'Because that's where the action is.'

"Girls," Augie laughed. He threw up his hands. 'Oh, boy, that's all right, I guess, if that's what you want to do." Augie never married, although he says he has a "gal friend."

In the 1960s, Augie quit growing strawberries. The canneries moved to the Kent Valley and Oregon, increasing the challenges of getting his fruit to market. Health department regulations continued to raise his expenses. The labor force dwindled. In later years, he planted holly to supplement the slow-income winter season, but by the time his trees grew ready for harvest, customers preferred plastic wreaths and berries. Years later, in the early '70s, the market changed and real holly wreaths became popular again, so Augie sold holly by the pound to wreath-makers who traveled from Silverdale to harvest it from his farm.

"Don't think I'm getting rich off this stuff. To clean these fields I have Harold Mann [see Chapter 1] come in and mow it and keep it so I can work in there. We fertilize the trees and keep the brush from taking over. I break even; that's good enough."

In about 1974, Augie planted Christmas trees, hoping that would be a way to make a living off the land, supplemented by a few raspberries, loganberries, and strawberries. For several years, he maintained a U-pick farm where he lives on SW Bank Road, but he quit that in the late 1990s. Today during the winter season he can be found at his Christmas tree farm, Augie's U-Cut; he still sells Christmas trees and holly wreaths. He leases the land to Roger Sherman and Carl Olsen; Augie is mostly retired.

When asked what advice he'd have for new farmers, Augie says, "Anybody who is going to try farming now, I would suggest they have a very good independent income and treat farming as a hobby. I don't think these small places can make a living for a family—I don't think you can do it.

"But Vashon is a special place," continues Augie, who recently attended his high school reunion at the [Vashon Island Golf and] Country Club, getting reacquainted with old friends here. "I'm glad that I farmed. I've enjoyed farming all along."

Augie's U-Cut [Christmas tree farm], 11003 SW Bank Road, (206) 463-2735

A Brief Time Line of World War II

September 1, 1939—World War II began with Germany's invasion of Poland, and the conflict expanded to include most of the countries in the world.

December 7, 1941—Japan bombed the U.S. Pacific fleet at Hawaii's Pearl Harbor.

February 19, 1942—Executive Order 9066 was issued, affecting all Japanese Americans living west of the Columbia River. Every Japanese family on Vashon (80 to

100 people) were given two weeks' notice that they would be interned. More than two-thirds of the 112,000 interned were Nisei (second-generation)—and American citizens; more than half of them were infants or children.

During the war, several thousand volunteer Japanese-Americans served in the all-Nisei 442nd Regimental Combat Team. Together with the 100th Infantry Battalion, they became the most highly decorated military unit in U.S. history. Nisei women also served in the Women's Auxiliary Corps (WAC's).

By the end of 1942, challenges to the constitutionality of the internment made their way to the U.S. Supreme Court.

December 17, 1944—The U.S. government, fearing a negative court decision, announced the end of the mass exclusion order against Japanese-Americans.

December 18, 1944—The U.S. Supreme Court ruled that the U.S. government could no longer detain loyal citizens.

August 6, 1945—United States bombed Hiroshima, Japan.

August 9, 1945—United States bombed Nagasaki, Japan.

May 8, 1945—Germany surrendered.

August 14, 1945—Japan surrendered, bringing an end to World War II.

March 20, 1946—The last of the ten major detention camps, Tule Lake, was closed.

Chapter 6

Mary Matsuda Gruenewald: The Strength of Family

The Matsuda family drew strength by maintaining strong family ties. Pictured here are Heisuke and Mitsuno Matsuda and their children Mary (Matsuda) Gruenewald and Yoneichi Matsuda. Photo courtesy of Mary (Matsuda) Gruenewald.

Heisuke and his son Yoneichi Matsuda. Photo courtesy of of Vashon-Maury Island Heritage Association.

Heisuke Matsuda was born in Japan. As a bachelor, he was a resourceful man who saved the money earned in America. When he decided to get married, he returned to Japan seeking a bride who was willing to come to America. Through the customary practice of using go-betweens, he met and married Mitsuno Horiye. The young couple came to America through Seattle on March 29, 1922, two years before laws were passed that restricted Japanese immigration to the United States for more than two decades, creating an age gap of an entire generation among Japanese immigrants.

First-generation immigrants, including Heisuke and Mitsuno, are called Issei. Second-generation Japanese, including Mary and her brother Yoneichi, eighteen months her senior, are known as Nisei. The third generation, including Mary's children, are called Sansei.

Heisuke and Mitsuno lived in Fife and Seattle before leasing a 10-acre parcel in the Shawnee area of Vashon Island and moving there in 1926 with their two children. Mary, who was two years old at the time, remembers the "big drafty old house" with the vegetable garden, sour cherry trees, and Italian plum tree. When the plums ripened, her mother placed a white sheet on the ground to catch the plums that fell when her father climbed up and shook them loose. "Oh, how sweet and delicious they were," Mary recalls.

The property had a barn where the family kept a work horse named Dolly. Their farm, which later included loganberries and cherries, attracted out-of-work drifters feeling the oncoming effects of the Depression, who slept on hay in the barn by night and worked the fields by day. Water was pumped from a nearby creek and stored in a tower, then gravity-fed to the house.

It was a productive family farm. Her father hitched Dolly to the wagon, loaded the vegetables into large gunnysacks, and headed for the Burton or Shawnee dock, then loaded the produce onto the passenger boat and headed for Tacoma. When they docked at Tacoma, the bags of peas or cabbage were lugged up a long stairway by a vendor who would sell the crops at the market. Her father also sold produce to the store at Magnolia Beach on Vashon.

Back then in about 1929–1930, sour cherries sold at one half cent to one cent a pound and earned a good profit. In those days, one could buy 10 acres of land and build a house and a barn for $2,000. Her father saved his money in hopes of buying his own place. But Japanese-born people were prohibited from owning land by the 1921 Alien Land Law (see sidebar "U.S. Laws Affecting Japanese and Japanese-Americans"). In 1930, Mary's father bought 10 acres of land at Center, purchased in the name of the son of a Japanese-American family friend, to be held in trust until their first-born son, Yoneichi, came of age. The acreage was behind the K-2 factory in the central part of Vashon Island. There, Heisuke had a home built, with four bedrooms, indoor plumbing, and hot and cold running water in the bathroom.

Heisuke was an enterprising man who experimented with crops and grew loganberries, Olympic berries, gooseberries, and eventually currants. He designed a system to increase the efficiency of the gooseberry harvesting process, by constructing flats, screens, troughs, and a giant fan to separate the leaves from the berries.

Loganberry growing is heavy, intensive work. The canes grow along the ground in the winter months. Then in the spring, workers wearing heavy gloves lift the vines and weave them along parallel horizontal wires to fill the spaces between the wires. Mary remembers sitting beside her parents at night with a needle, picking the thorns out of their fingers.

Currants were easier to harvest, because they are bushy and grow like clusters of grapes. The flats could be stacked and packed to stations close to where the pickers worked.

The work ethic permeated their whole family, Mary says. "We had a woodstove. When these big pieces of wood [indicates 10–12 inches in girth] were cut into stove lengths, Dad would break them with wedges and whack them into smaller ones. Then my brother and mother would cut them into the size for the kitchen stove. And then I would come along with my little hatchet, and I would make kindling. We always did things together," she remembers.

Her father always thought about the future, how to improve things, how to make things better. Her mother was a nurturer, a spiritual person who appreciated the simple pleasures of life, Mary recalls.

"On a hot August day, when we were so hot, hunched over picking currants, a breeze would come along," Mary remembers, "and Mother would say, 'Ah! Isn't that breeze wonderful!'"

"There was always an air of appreciation and thankfulness," Mary says. "When we worked together and we each did our part, it made it go better for all of us."

Native Americans helped the Matsuda family harvest their crops for several years in the early 1930s. Mary recalls that they pitched tents in the yard and caught salmon nearby, filleted them, and set them out to dry in the sun.

As the farm expanded to the west, her father sold crops to Seattle (including Borden's), Tacoma, and Puyallup (including Hunts' Brothers), sending them via Vashon Auto Freight. In the late 1930s, when Masa Mukai set up his berry processing plant later known as VIPCO (see Chapter 7), the Matsudas sold berries to him.

In the years before World War II, Heisuke hired families from Vashon and the mainland to harvest his crops, working alongside them, encouraging them, and tutoring them about how to pick the best fruit. "My dad always had a three-pound coffee can filled with candy," Mary remembers. "He bought big bags of butterscotch, peppermints, tootsie rolls—all kinds of good stuff. At 10 every morning, Dad would go around and give candy to all the pickers. It was hard work for those kids. Candy gave them something to look forward to—it's a reward for doing a good job.

"In between all this, he and my brother would squat down beside the different pickers, and they would pick berries and put them into the kids' carriers—and the kids loved it because their carriers would get filled up faster. But in the process, Dad and Yoneichi would teach: 'You pick all those ripe ones. You don't put green ones in, and you don't pick the big ones from the next person's rows. You pick it clean. And you don't put rocks and leaves on the bottom of the box. You go with the best berries, because when they go to the market, people want good, ripe strawberries.' It's that kind of work ethic," Mary says.

"So kids would come, and over time then their brothers and sisters, and sometimes their families. Some whole families came through and earned money for their school clothes. No matter how young they were, we let them pick berries if their parents came with them," she recalls.

Her father recruited youth in Seattle, and in the summers fourteen or fifteen boys bunked in their barn from late May through mid-August, during the strawberry and currant seasons. The kids were awakened at 5:00 a.m. and worked until 3:30 p.m. Mitsuno fixed their meals, serving them on a long picniclike table with fresh fruits and vegetables from their garden—apples, peaches, nectarines, grapes, figs, fresh lettuce, peas, carrots, spinach, salad greens.

Those kids developed strong, healthy bodies and saved their money, Mary recalls. At night they played baseball and basketball in the yard. "That's the way the labor force went, gradually, into the 1940s," Mary remembers. Her family was among those who were interned during World War II (see "A Personal Account of the Internment of Japanese-Americans" at the end of this chapter).

After the war, Mary's parents returned to Vashon, where her father continued to farm until he was injured by his horse in the spring of 1946. After his father's injury, Yoneichi was sent home on emergency leave from military service to carry on the family farming business. Yoneichi attended college, then taught at Ingraham High School in Seattle, and continued to manage the farm, growing strawberries and currants, expanding it to 50 acres. He retired from teaching in June 1985, sold off part of the land, and died in September of that year.

Mary went to work for Providence Hospital and attended the College of Puget Sound in Tacoma, then the University of Washington in Seattle, and earned a certificate in public health nursing. She married Charles Gruenewald in 1951, joined him in Boston, then returned to the Puget Sound area in 1953.

Her mother and father died in 1965 and 1975, respectively. The land is still in the Matsuda family, with 27 acres remaining, and is harvested for hay. Miyoko Matsuda, Yoneichi's second wife, resides on the property. (Yoneichi had been previously married to Marjorie Nakagawa in 1958 until she died on September 7, 1973.)

A Personal Account of the Internment of Japanese-Americans

Mary Matsuda at age 17, when her family was taken to internment camps during World War II. Photo courtesy of Mary (Matsuda) Gruenewald.

December 7, 1941, was a Sunday. Mary and her brother, Yoneichi, were at the Methodist church on Vashon, dusting the pews and lining up the hymnals, as they always did. They were unaware of what had transpired in Hawaii on that fateful day that Pearl Harbor was bombed. As they were leaving the church, someone said to Yoneichi, "Well, your countrymen did their dirt."

"We didn't know what he meant. When we got home, my dad was in the house. My folks were real serious-looking," Mary recalls.

"Japan has bombed Pearl Harbor," he told us. Her heart sank.

"We knew immediately that there was going to be trouble for us," Mary says.

The next day she went to school as usual. She was a junior in high school, treasurer of the student body, member of the honor society, a Vashon Islander for eleven years—but she no longer felt a part of the class.

At home, the family gathered at the radio. Blackouts were ordered; each night the family covered the windows with blankets. Curfews required all Japanese Americans to be inside their house after 8:00 p.m. Their bank accounts were immediately frozen.

A family friend who was president of the Japanese Association was taken by authorities. Mary's father was secretary-treasurer for that association. "The night that Mr. Usui was taken, I knew that Dad was at risk," Mary remembers.

A friend from Bainbridge Island phoned to tell the Matsudas there had been a raid, and urged them to get rid of everything that would tie them to Japan. "We pulled out pictures, phonograph records, dolls—exquisite dolls—books, and we burned everything. We were not rational—we were terrified," Mary recalls. "We burned all the stuff, hoping Dad would be spared."

One Saturday morning, two tall, slender men wearing black suits and hats knocked on their door. Agents of the Federal Bureau of Investigation, they said, "Is this the home of Heisuke Matsuda? We would like to talk to him."

Mary, who had been cleaning the breakfast dishes, ran to the field to tell her parents. Her brother and father accompanied her to the house.

The FBI agents looked around, taking it all in. Looking at a broom closet, one said, "What's in there?"

Mary opened the closet and showed them her brother's 22-caliber rifle. "He uses it to shoot the crows that eat our strawberries," she recalls telling them.

Moving into the living room, one of the agents motioned to the console radio: "Can you get Japan on this?"

Her brother replied: "Yes, once in a while in the evening when the weather conditions are good."

"How do you transmit?" one asked.

"We don't do that, we don't have the equipment," he replied.

They continued to search the house, then thanked the family and left.

Some of the twenty or twenty-five other Japanese-American families on Vashon Island experienced far more invasive searches. "Some of the agents put their hands through rice and sugar, looked behind furniture, and rifled through drawers," Mary remembers.

Signs posted around the island and leaflets that were circulated said, "All persons of Japanese ancestry, both aliens and nonaliens, will be evacuated by 10:00 a.m. on May 16, 1942." The term "nonaliens" referred to U.S. citizens of Japanese descent.

"It is unconstitutional to imprison U.S. citizens, but if they called us 'nonaliens', a strange euphemism, they could get around the law. Our parents were foreign-born, and by law they couldn't become citizens; they were considered alien," Mary explains.

Her family was directed to report to a location about a mile north of Center.

"We walked. As we approached, we saw army trucks and soldiers with rifles and bayonets, standing at attention. We said, 'Oh!' in terror. We were afraid they were going to shoot us. They were using the Army to evacuate us, but my dread throughout the whole camp experience was that they were going to kill us.

"We piled into covered army trucks and were driven down to the ferry. Not the usual Vashon ferry. It was another one, smaller. The dock was lined with neighbors and friends—300 or so, it said in the *Beachcomber,* Vashon Island's newspaper. I couldn't really see; I was teary-eyed. We all walked onto the ferry, and the soldiers went with us. I'll never forget the moment when we pulled away from the dock. The people on the dock waved. One of my friends was there waving, waving, waving. I could still see her waving, waving as we moved out farther and farther away from the Vashon dock."

In Seattle they trudged to a waiting train. Men wearing coveralls and holding shotguns yelled down from an overpass, "Get out of here, you goddamn Japs! I oughtta blast your heads off!" Others spat at them.

It was a dirty, old, rickety passenger train. They crowded on, packed beyond capacity. They took turns sitting down. Mary describes herself as an immature seventeen-year-old at the time. She asked her dad, "We're all going to go to the same place, aren't we? Can we all stay together?"

"I don't know," he replied.

"It was chilling," Mary recalls. "He didn't know! Dad always knew the answers."

They rode for days in the filthy train, with smoked windows to hide them from the people of the towns they passed through. There was no fresh air. In the 100-degree heat, Mary fainted. When the train stopped, somebody barked, "Everybody off." Next they boarded army trucks. When the trucks stopped, a soldier shouted, "Everybody off. End of the line!" They put a stool down for people to step down onto.

Mary looked up. They were inside a compound, a prisoners' camp, surrounded with high fences with electrified razor-blade barbed wire all along the top. Huge searchlights revolved 360 degrees, twenty-four hours a day. Guard towers with soldiers armed with machine guns were strategically located around the camp. Mary thought to herself, "I deserve this?" and fell into a deep, discouraged depression.

Their first meal was hot white rice and chili, along with a salad of cabbage, lettuce, slivers of radishes, and shredded carrots, which had been reduced in volume by vinegar and hot water. The meal was served in a massive military mess hall with picnic tables and a noise Mary describes as horrendous. She was too heartsick to eat, though she chewed and chewed.

Her mother tried to help. "It's hard to eat, isn't it?" she said. "But it may be all we get today. So try to eat what you can."

"My mother had the capacity to be outside of herself in the situation," Mary remembers, "to figure out what to say so that the situation was tolerable, manageable, no matter what she might have thought."

At the camp, they occasionally saw some of the other Vashon farmers—Augie Takatsuka (see Chapter 5), Tok Otsuka, and the Yoshida family.

After six weeks at the Pinedale Assembly Center, eight miles north of Fresno, California, her family was told to pack their things and report to the gate by 7:00 a.m. They were transported to Tule Lake, in northern California near the Oregon border. The camp was in a huge, craterlike area that looked like a former lakebed.

At the camp, Mary tried to make sense of life by listening to an evangelical bible station called "Voice of Prophecy." She underlined passages in her bible, looking for answers. Her parents noticed her withdrawal, and her mother took her for a walk to find some shells. That walk became a daily ritual to maintain their sanity. Mary cleaned the shells and lacquered them with fingernail polish purchased at the canteen, then strung necklaces and made brooches and wall hangings to mail to friends.

In September 1942, Mary attended school in camp. The teacher drew a typewriter keyboard on a large piece of butcher paper, and the students typed on pretend keyboards on their laps. The teacher had the only textbook. The students learned chemistry without a lab or any equipment.

The government developed a document of forty questions that was designed to separate the loyals from the nonloyals, Mary recalls. Everyone over seventeen years old was required to answer it. Among the questions were: "Are you willing to serve in the armed forces of the United States, in combat duty, wherever ordered? Will you swear unqualified allegiance to the United States and faithfully defend the United States from any and all attack by foreign or domestic forces, and foreswear any form of allegiance or obedience to the Japanese emperor, or any foreign government, power, or organization?"

By the following February, the camp residents grew agitated and demanded their rights. Violence erupted. There were beatings. The Army came in with machine guns, surrounding the barracks and taking the agitators to the stockade. There were hunger strikes. Mary's family held a meeting in their cubicle, whispering for privacy, to decide what to do. Her brother chose to fight in the war to prove his family's loyalty to the United States. They all voted yes, yes.

"I realized there are times when the country will not do for you what you think a country should. And so I realized at that moment that the sense of family—the containment, the safety, and the love of the family, no matter what happened to us in the future—the family meant all."

Her family boarded a train and transferred together to Heart Mountain, Wyoming, and stayed there for a year. From there, her brother enlisted in the 442nd unit of the Nisei Regimental Combat Team and fought with it in Italy. Mary was accepted into the U.S. Cadet Nurse Corps, and reported to training and work in a 100-bed hospital in Clinton, Iowa.

Looking back on those difficult times, Mary holds a remarkable lack of resentment for how she and her family were treated.

"I think bitterness is corrosive to the person who holds it," she says of the internment experience. "It does nothing to solve whatever the problem was that caused it in the first place. Now that doesn't mean I blame anyone for being bitter. There are some people who still are. We all do and say things—acts and sins of omission and sins of commission. And because I intentionally or otherwise hurt people, I have to give that same right to other people.

"Forgiveness is the basis for a rich life. We always have the option of choice. I think each one of us has the opportunity to create or choose how we look at or come through a difficult time. It is in the process of sharing, out of our abundance of love and forgiveness, that our own richness comes back," she says.

Mary Gruenewald is writing a book about her wartime experiences. She lives in Seattle, Washington.

U.S. Laws Affecting Japanese and Japanese-Americans

Compiled by Mary J. Matthews

Perhaps the most significant challenge to Japanese people in the United States were laws that prohibited the Japanese from buying land. This restriction reduced first-generation immigrants to tenant status virtually in perpetuity, and was the major cause of Japanese Americans not returning to their farms on Vashon after World War II. Land ownership was denied the Japanese: The Washington State Constitution restricted ownership of land to those immigrants who declared the intention to become U.S. citizens; and the Constitution of the United States restricted citizenship to only those individuals who qualified as a "free white person."

Citizenship was extended to persons of African descent after the Civil War, but discrimination as denial of citizenship rights toward Asians continued with the 1882 Chinese Exclusion Act. In 1906 U.S. courts were specifically ordered to stop granting citizenship rights to

Japanese, and this order was reaffirmed in a 1922 U.S. Supreme Court decision.

The 1921 Alien Land Law, passed by the Washington State Legislature soon after a similar law was passed in California, extended ownership restrictions to prohibit Asians from leasing or renting land. It was not until 1952, with the passage of the McCarran-Walter Act, that Japanese immigrants were allowed naturalization.

Executive Order 9066, issued on February 19, 1942, affected all Japanese Americans living west of the Columbia River. Every Japanese family on Vashon (between 80 and 100 people) were given two weeks' notice that they would be interned. On December 18, 1944 the U.S. Supreme Court ruled that the U.S. government could no longer detain loyal citizens. World War II had a lasting effect on Vashon's Japanese agricultural community. Those who were leasing at the time of the internment lost virtually everything, and more than one-third of the evacuated Japanese never returned to the Puget Sound region.

Chapter 7

Masahiro Mukai's Vashon Island Packing Company: The Vashon Island Strawberry Industry

VIPCO, an innovative berry freezing plant owned by Masahiro Mukai, changed the strawberry-growing industry on Vashon Island. Photo courtesy Vashon-Maury Island Heritage Association.

Strawberry fields were forever fruitful at the Mukai field. Photo courtesy Vashon-Maury Island Heritage Association.

By Mary J. Matthews

The Mukai Farm and Garden, located in the center of Vashon Island, is a 2.75-acre site remaining of what was originally part of a 40-acre farm purchased by second-generation Japanese-American Masahiro Mukai in 1927, when he was 16 years old. It is probably the most significant historic site in the Puget Sound region associated with first-generation Japanese immigration, settlement, agricultural development, commerce, and, on a private level, the art of the garden and the role played by Issei women. The Mukai Farm was designated a King County landmark in 1993, the first such property associated with Asian-American history in King County; and in 1994 it was listed on the National Register of Historic Places. The Mukai Farm and Garden achieved this status for two reasons:

First, the traditional Japanese "stroll garden" that surrounds the house was designed by Issei woman Kuni Mukai. In the Meiji-era Japan of Kuni's youth, men, not women, designed gardens. Therefore the garden is more than an exquisite work of art; it is a fascinating cultural artifact, interpreting what happened to one Japanese woman, representative of many such women who broke from their traditional roles upon immigration to America.

Second, Mukai Farm and Garden played a pivotal role in Vashon Island's strawberry-growing industry. The enterprising Mukai family was the first in the region to develop a technique for freezing strawberries, which enabled them to escape from the financial restrictions of the fresh-fruit market and expand from local markets to those around the world. The family's strawberry packing plant also provided a crucial local buyer for many island strawberry farmers who were faced with multiple challenges marketing their crops.

It all started around 1902, when B. D. Mukai jumped ship in San Francisco Bay at the age of sixteen and unofficially immigrated to America. Believed to have been from a family of samurai lineage, Mukai was resourceful and hard-working. After learning English and working as a houseboy, police interpreter, and owner of an employment agency, Mukai and his wife, Sato Nakanishi, moved to Seattle after the San Francisco earthquake in 1907. A job with the Walter Bowen Company on Wholesale Row introduced him to Vashon Island's Marshall strawberries, a large, bright red variety.

The strawberry industry was established on Vashon Island by Caucasian growers as early as 1883, in small family plots. Most growers around the turn of the century were growing the soft and sour strawberry of the Magoon variety. By 1901 more than 15,000 crates of strawberries were shipped annually from the island, and it was also the first year of the Vashon Island Strawberry Festival, a long-standing tradition that continues to this day.

At the time of B. D. and Sato Mukai's arrival on the island in 1910, strawberries were already a major Puget Sound crop. Vashon Island soil is glacial and, though not particularly suited for high yield, excellent for early yield. In the fresh-fruit market, early berries commanded the best prices in Seattle, giving Vashon the edge over larger-producing areas such as Bellevue and the White and Green River Valleys.

It was the Japanese immigrants, with their cultural history of working the land and their remarkable endurance, who made strawberry growing their own. These immigrants usually had almost no capital and so rented or leased land for a part of the future crop, commonly known as sharecropping. Five to 10 acres was the common leasehold. Leases normally ran from one to eight years, which would cover one or two crops because a strawberry plant lives for five years: No crop is produced in the first year, and the second, third, and fourth years are the best producing years. Because of the depletion of nutrients in the soil, new fields have to be planted in the fifth year.

Around 1910, a strawberry-growing season consisted of several months of idleness (October through March) contrasted with several months of hard work (April through September). April through June was the busiest time, with harvest occurring in late May and early June. During the picking season, six to ten pickers per acre worked from dawn to dusk to bring in the harvest, because the fruit is highly perishable. Both Caucasians and Japanese worked as pickers in the early years of the industry; later, Native American women and children provided the necessary labor.

In 1910 B. D. and Sato Mukai started growing strawberries on a small parcel of land they leased near the center of the island. When Mukai began his strawberry farming, he was a tenant farmer, and he hired other Japanese to pick his crop. After a year on their first farm, and after the birth of their son Masahiro in 1911, the Mukais moved to the Cove area. Two more moves brought them to a 60-acre farm owned by the Taylor brothers. It was there, in 1924, that the family began raising strawberries for the frozen, rather than the fresh, market.

When the Mukais began cultivation of the Taylor brothers' site, the strawberry industry was booming. The value of the strawberry crop exceeded that of any small fruit in the Puget Sound region, with 39 percent of fruit farm receipts coming from strawberries.

Working with "Dutch" Diehl, a U.S. Department of Agriculture agent, B. D. and Masahiro Mukai set up the first barrelling of strawberries in a shed out in the field. The berries were picked and then packed in wooden barrels made from Douglas fir staves shipped in from Port Angeles. Each barrel held 300 pounds of strawberries mixed with 150 pounds of sugar; when packed, the barrels weighed 450 pounds net and were marked "2+1 Marshall Strawberries," referring to the proportion of berries to sugar. The barrels were rolled onto horse-drawn drays and delivered by the next morning, at the latest, to Seattle's Spokane Street Cold Storage, where they were frozen. They were then shipped all over the world.

The Mukais' 60 acres yielded 200 tons of strawberries, and 1925 was a peak year: Weather allowed them to produce a double crop, in both June and October. They made $70,000 selling strawberries, most of them served in confectionaries in London, England.

Having established a new market for strawberries, the Mukais used this money to buy a 40-acre farm, purchased officially by sixteen-year old Masahiro, who as a U.S. citizen could legally hold title to land. On their new farm, they built a house, an extensive garden, and a large fruit barrelling plant. With this plant, the family was able to compete with the Puyallup and Sumner Berry Growers Association for Vashon Island farmers' strawberry crop.

At this time, Marshall strawberries were in great demand by large distributing companies such as John L. Sexton Company and the J. H. Smucker Company. The founder of the Smucker Company visited the Mukai farm in the 1920s to see its Mukai Cold Process Fruit Barrelling Plant.

The Mukai farm prospered until 1938, when anti-Japanese sentiment forced Masahiro (known as Masa) to rename the business Vashon Island Packing Company, or VIPCO, as it was more commonly called. The office was moved out of the house to a new brick building, constructed on what was originally a part of the stroll garden. When the war broke out, Masa knew a high-ranking official so the family was allowed to leave voluntarily. He left the business in the hands of his hauler, Morris Dunsford, and provided funds for its operation, then the family packed their belongings and moved to the Snake River Valley, Idaho, to live near a relative during the war years.

Executive Order 9066, issued on February 19, 1942, affected all Japanese Americans living west of the Columbia River. Every Japanese family on Vashon (between 80 and 100 people), except the Mukais, were given two weeks' notice that they would be interned. Most of the island families were moved to the Puyallup Fairgrounds and then the families were dispersed to northern and central California, Wyoming, Idaho, and other locations. More than two-thirds of those interned were Nisei (second generation)—and American citizens. Those who were leasing at the time of the internment lost virtually everything, and more than one-third of the evacuated Japanese never returned to the Puget Sound region.

When the Mukai family returned to their farm after the war, they found the farm and the fruit packing industry greatly altered. Their house was rented and the garden had been neglected. Masa moved his family to Seattle and commuted to Vashon for a time. In 1946, he planted a 60-acre plot north of the town of Vashon, but he found that it was no longer profitable due to the ferry cost and the difficulty in securing laborers and pickers. Abandoning his family's longtime agricultural business, Masa continued to operate the VIPCO plant and purchase berries from other island growers. It was not until 1969 that he gave up the struggle and the plant was sold, putting an end to Vashon Island's strawberry packing and freezing business.

A few growers, notably Tok Otsuka and August Takatsuka (see Chapter 5), struggled on. In 1967, the last year that Tok Otsuka grew berries on the island for the commercial market, marked an official end to the once-great and thriving industry. Tok still sells corn in August, and Christmas trees in December, from his farm at 16917 Vashon Highway SW.

Masa Mukai died in 1999. The Mukai Farm and Garden, located on SW Bank Road just west of Vashon Highway SW, is presently in a planning stage and is not open to the public except by special appointment.

The Mukai Farm and Garden, 18017 107th Avenue SW. For information, call Jon Knudson at (206) 463-2349

Part III
TREE FRUIT ORCHARDS BRANCH OUT

Midcentury Food Producers from the 1950s On

Chapter 8
Betsy and Robert Sestrap: Wax Orchards Farm

Betsy (Wax) Sestrap gathers peaches at Wax Orchards Farm. Photo courtesy Vashon-Maury Island Heritage Association.

Astute attention to world trade trends, weather patterns, and marketing niches has been the secret ingredient in Wax Orchards Farm's recipe for success. August Wax and his wife, Johanna, began their chicken farming enterprise in Covington, six miles east of Kent, Washington, in 1920. At the family's peak, they raised 40,000 laying hens among four locations—Covington, Panther Lake near Renton, Kent, and Vashon Island.

On Vashon, Lewis Beall's (see Chapter 4) pedigreed, blue-ribbon, prized white leghorn roosters had sired the Wax family's chicken flocks, and Beall is part of what eventually brought the Waxes to Vashon. Cherry farming in Covington was added to their enterprises, from rootstock imported from France. Growing cherries appeared to hold the promise of a less labor-intensive and more lucrative endeavor than the year-round task of chicken-rearing. Unfortunately, the Covington location was in a frost pocket where the fruit didn't thrive.

August Wax found a perfect microclimate on a piece of land on Vashon Island 3 miles north of Tahlequah, where the weather rarely froze. In 1929 he bought 60 acres there. He sent Johanna and daughters Betsy and Linda ahead to start the family's Vashon chicken ranch and establish a cherry orchard. "We had raised enough cherries in Kent to know that they would be less troublesome than chickens," recalls Betsy from her Wax Orchards Farm home. "Fruit-raising looked to be a vacation," she says.

In the early 1930s, the family built two 250-foot chicken houses on Vashon and raised the delicate leghorns from chicks. Why raise leghorns if they are so delicate? "Because they are egg-laying machines!" Betsy replies emphatically.

Betsy relates an incident she witnessed as a fifth-grader that gives a glimpse into the true grit and determination her mother brought to chicken-rearing. Workers repairing their chicken house dropped a box of roofing nails onto the ground. The chickens descended on them. "A chicken will eat anything," says Betsy, "and the darn things ate those nails."

The nails lodged in the pullets' crops, the fold of skin that hangs in front of their throat. "My mother opened up the crops, took out the nails, and sewed them right up again," Betsy remembers. "It didn't bother her a bit," laughs Betsy. "The women in this family are no-nonsense."

A smart and tough businesswoman, Betsy is also quick to give credit to the others involved in the enterprise. Betsy's dad maintained the profitability of his chicken business by keeping abreast of world trends. He tracked shipping and trade statistics, bought his feed wisely, and bypassed middlemen by building his own feed grinding mill. The family shipped their eggs to New York to brokers. But no matter how wise a businessperson a chicken farmer is, the fact is that the chicken business is high maintenance.

"Leghorns are nervous. If you walk into the henhouse too quickly, they fly up and get upset. You can't let them get chilled. You can't let the water troughs go dry. You have to be watchful or they won't lay eggs," Betsy says.

The family eventually quit raising chickens when the trends turned to caging them and increasing their daylight hours with batteries and artificial lights. Betsy says, "We didn't want to go that way."

When the cherries came into fruition, so did the family's orchard enterprise. They discovered that the particular type of soil on Vashon's land, called glacial till, was perfect for growing a variety of fruits.

Their early labor force was comprised of Native Americans who traveled to Vashon from British Columbia to harvest the first fruit of the season—strawberries (see Part II, *Strawberry Fields Forever*). After that harvest season was finished, the Waxes gathered the workers from the strawberry camps and brought them to Wax Orchards Farm. Cherries were their first crops ready for harvest, followed by hops and raspberries, then apples, gooseberries, and currants. Providing several crops that could be harvested in succession was the key to retaining the labor force for the longest period of time. Workers numbered about 150 on the average during fruit-picking season. At their peak they numbered 400.

"We had a bumper crop one year. They were sleeping under the trees. They came from everywhere—they heard about the crop," recalls Betsy. The workers spent much of their income (one ticket per 3-gallon bucket of cherries) at the general store known as Cherrywood Station, managed by Charlie Mackie, who also owned a grocery store at Cove.

Betsy grew up and in 1947 married Robert Sestrap, a grain and cattle rancher from Alberta, Canada. August's wedding present to the couple was a 20-acre parcel on Vashon Island with an apple orchard. Over time, the family amassed 270 acres on Vashon. The Sestraps raised three children—Anna, August, and Kathryn—who "checked cherries" for their summer income. During the mid-'40s, Vashon youth provided the labor force, including one of the "kids from Cove," Tom Lorentzen (see Chapter 3). August Wax died in 1951.

The Sestraps experienced a setback in the 1950s when they took the advice of a university agriculture consultant, who told them that the now thirty-year-old trees were finished bearing fruit. "'Pull them out,' he said," Betsy recalls. So they removed 60 acres of trees.

On a nearby acreage, they cleared an orchard by "dehorning" the trees, an extreme pruning method that cuts the tree back to a few feet of branches up from the main trunk. Those trees began producing fruit two years later, and continued to produce for twenty more years, Betsy says. That's when the Sestraps realized they had lost thirty years of crops and income from those 60 acres.

"We lost millions of dollars," Betsy says. "To this day, we feel betrayed." When they discovered other orchard businesses were acting on the same bad advice, they formed the Tree Fruit Association to spread knowledge among the growers.

A hailstorm in the early 1970s ruined their Gravenstein apple crop, bruising it so badly the apples couldn't be sold fresh. Betsy and Robert used an old wine press and ground the apples into cider. That experience provided their initiation into the fruit processing business. Three varieties of apples were ground together into pulp, blending the best qualities of each. Initially the fruit was shipped to Seattle to be frozen; then the Sestraps added a 3,000-square-foot freezer to their operations. In the early 1970s, pickers were recruited from the mainland and ferried over during the summer months.

In the 1980s, consumer demand for natural fruit sugars grew. Betsy concocted formulas and recipes, and created a line of tasty "fruit-sweetened" products. Robert invented and custom-built factory equipment that could give a small family business an edge over similar-sized enterprises. Robert has invented more than seventy kinds of machinery, controls, and systems that are in worldwide use today.

In 1986, the year that Johanna Wax died, a *Dallas Times Herald* article about Wax Orchards Farm's products hit the news wire services, and consumer interest skyrocketed. Betsy recognized that their product niche was in the "fancy food" industry. Then their products were picked up by institutions serving patients seeking dietary alternatives to refined sugar. The business has grown rapidly in recent years, gaining more than half their customers in the last couple years, and they now have their own web page.

In August 2000 the Sestraps sold 140 acres of their land, including the Wax Orchards Airport, which had for years been designated open farmlands, to neighbor Tom Stewart, owner of Misty Isle Farms, where Black Angus cattle graze on the green, rolling pastures. (Tom Stewart declined to be interviewed for this project.) The Wax Orchards land had for years been enrolled in the King County Farmlands Preservation Program, in which the county purchased the development rights to the property to ensure it would remain available for agricultural use. Stewart's ownership of the former Wax Orchards land is bound by the same development restrictions. Now amassing a total 230–250 acres under the names of various corporations and ventures, Tom Stewart is one of the largest landowners on Vashon.

The Sestraps retained the 15 acres that include their home and the 12,000-square-foot cannery. The rest of the acreage was divided among family members. One of the Sestrap's children, Anna Sestrap Swain, continues in the family business as their financial officer; August Sestrap lives on Vashon Island, and Kathryn Sestrap lives in Seattle.

Betsy says they have no plans to retire, and the proceeds from selling their land will be used to invest in marketing their food-processing business. They'll hire more crew to manage their thirty-one products and gain more free time for themselves.

If she had it to do over again, would she still farm? "In farming, you don't really have choices," Betsy replies. "We've never been faced with choices. We've been faced with opportunities that we grasped," she says.

Wax Orchards Natural Products, 22744 Wax Orchards Road SW, (206) 463-9735, www.WaxOrchards.com

Chapter 9
Pete Svinth: Pete's Farm

Pete Svinth digs his trees, and left a legacy of propagated trees when he moved away from Vashon Island. Photo © Ray Pfortner.

One of the best-known Vashon growers is Pete Svinth, founder of Pete's Farm, better known as Peter's Pumpkin Patch, which was actually on Maury Island, a part of Vashon Island. His pumpkin patch was a regular stop for young kids, who delighted in sprinting around October's soggy fields in search of their own live pumpkin for a Halloween jack-o'-lantern. His casual expanse of farm was a favorite destination for locals and mainland visitors of all ages for two decades. Pete's slow, to-the-point speech and ambling gait were a welcome respite from the fast pace of everyday life.

"The pumpkin patch was an outreach to the children," Pete says, gesturing with his huge, work-worn hands. "When the children would come for the pumpkins, they could also pick an apple or raspberries and do these other things. I did not encourage tourists to come to the pumpkin patch. It was supposed to be a farm experience; it was supposed to be educational. The busloads of tour groups came only if I could meet with them and talk about the sex life of a walnut," Pete says. "The senior groups got a kick out of that."

On the Svinth side of his family (pronounced "Swent"), Pete's grandmother was of the Nisqually Indian Tribe, his grandfather hailed from Denmark, and Pete's mother and father were English-Welsh and Danish, respectively. That heritage contributed to the tenacious follow-through and love of land that locals associate with Pete.

Although Pete comes across as casual about growing, propagating, and business, he is well educated, both formally and informally. He took some agriculture courses in high school, graduating in 1949, then in correspondence courses during his three years with the U.S. Marine Corps he learned about soils; later he took a plant propagation course at South Seattle

Community College. He earned a bachelor of arts in business administration with honors from the University of Puget Sound in Tacoma in 1963, and stopped just short of completing his course work for a master's degree in business administration in economics at Gonzaga University in Spokane in 1968.

Pete had planted three orchards before he moved to Vashon Island in 1973. At first his Vashon farm was 8 acres, where he raised hogs, chickens, beef cattle, and goats and always kept a garden. Eventually he became most interested in propagating tree fruits. Ten of the eventual 18 acres of Pete's farm had been a filbert orchard, planted in 1939 by the famed Shane brothers. The Shane brothers were known for their propagation of nut trees, and many of the trees they planted still stand today, all over Vashon Island. In the early years, Pete's growing enterprise held nonprofit status, but the paperwork was unwieldy so he changed the designation, still essentially running it as a nonprofit, using the money earned to buy more seed and plant stock and giving back to the land. During its life, the farm went through many incarnations and names, including Pete's Farm (what all the locals called it no matter what name changes occurred), Peter's Pumpkin Patch, Tree Fruit Research Foundation, Family Trees Orchard and Nursery, and Family Trees and Gardens.

In 1976 Pete took early retirement from a management position with Pacific Telephone and Telegraph (later USWest and now Qwest), and his farming endeavors were supplemented by his retirement income. "You don't do my type of farming to get rich," he says. "You do my type of farming for the lifestyle and the benefits you get from being around the plants. I never know what I'm going to grow. I don't plan; I just do. And that's not businesslike. By design I don't want to be businesslike. Quite simply, I grew what I wanted. People came and harvested it and left me enough money so I could plant some more and grow some more. The biggest enjoyment was seeing them enjoy the place and realize how serene and wonderful it was to be around there," he says.

Pete's passion for tree growing was born when he attempted to graft a Stella cherry onto a wild seedling that grew in his yard. The tree responded by sending out five branches of vegetative growth that headed upward like water sprouts. Then he tried bud grafting, placing a Stella cherry bud on each of the five whips. He cut them off the following spring and within three years had a bearing cherry tree. That tree, which seven years later began producing 400 pounds of cherries a year, still stands.

During Pete's tenure as a farmer, he found the expertise provided by the state and county extension services very helpful, and he received training as a master gardener. He also utilized the tax savings benefits available through the open spaces agricultural designation, which saved him money as he farmed.

Pete is quick to draw a distinction between farmers and growers. Farmers do it for the money, he says, and growers do it for the love of it. He declares, "I'm a grower." When asked what advice he has for prospective farmers, Pete quickly retorts, "Don't do it for the money." Toward the end of his Vashon farming years, his farm grossed about $56,000 a year, he says.

Pete was one of the primary builders of the original Vashon Island Growers Association (VIGA) produce stand in the late 1980s. Armed with a hammer, a ladder, stubborn determination, and some lumber, Pete and a handful of other volunteers cobbled together a wood and corrugated-plastic structure for farmers to sell their wares on the piece of land in downtown Vashon across from Bob's Bakery; later this spot became known as the Village Green (see Chapter 15).

Pete wanted to give away trees for free or for nominal amounts, digging them up with a bulldozer and handing them out to people at his farm, and sometimes at the Vashon Farmers Market. But sometimes people place less value on something that they get for free, so Pete asked them for a signed statement assuring proper stewardship. His self-serve honor-system produce stand, a fixture for more than ten years, was also evidence of his trust in the honesty of fellow produce-lovers.

How many trees were on his property? Pete says, "Hundreds and hundreds and hundreds. It depends on what you count as a tree. I had babies—I would line out fruit stock, at least 1,000 a year." He tested varieties and experimented with propagation of primarily apple and cherry trees. Over the years, Pete became an expert on grafting and propagation, and gave classes and workshops, sharing his knowledge, trying to reverse the trend of people doing it all wrong.

"The problem with conventional gardens—I tend to want to call it European gardening—is that it's not natural. People don't work with plants; they quite often try to make a plant something other than what it wants to be. An example is pruning, where the tendency is to cut off the apex bud on an apple tree and cause it to grow a whole bunch of water sprouts at the expense of fruiting, rather than let it form its natural shape, which is like a slightly rounded Christmas tree," he says.

"It's hard to get people to understand. Using pruning as an example again: people prune trees in the wintertime, but consequently those trees require more attention the following year. In actuality, trees of any kind require very little pruning. They need to be encouraged sometimes to grow in certain shapes, but you work *with* the tree instead of against it," he says. The time to prune a tree to reduce its size is the summer; winter pruning is recommended to generate new growth, he says.

Along with his scientific knowledge, Pete professes a kind of spiritual intuition about his trees and plants. "I communicate with all plants, I believe. It comes in the mind; it's just knowledge that's suddenly there. You're aware of the tree if it's in distress, you're aware if you do something right. When you're around the plant, you feel the plant and the plant feels you, and it's spiritual. The things I know fly against an awful lot of what we've been taught, and that has to do with our relationship with plants and with our relationship with the spirits," he says.

Besides plants, his most important spirit is that of Tacobet (pronounced Ta-co-buh)—the Nisqually name for what is known as Mount Rainier. Pete treks to the slopes of Tacobet as other people might go to a temple to worship, he says.

Pete espouses the merits of Native American traditions of growing plants, and in his later years came to more actively embrace their way of life. "The Native Americans worked with plants, either through manipulating fire or by taking advantage of burned-off places. And they knew that you could prune by simply breaking the branch in the right place and it would heal itself better than from any sharp cut. They encouraged plants they wanted to grow where they wanted them. And they didn't worry about long, neat rows.

"Well, my gardening on Vashon had rows by necessity, but you'd never see me run a string to make a straight row. And I would not use a seeder. Everything was done by hand. My concept of weeding is to reduce competition for the plant that you prefer to grow. My concept of insect control is the same. If the plant is healthy, it acts as its own insect repellent."

Does this mean he farms organically? Pete says he doesn't like the term, because it has become a commercialized concept. "The word I use is 'natural.' Be tolerant of insects, weeds."

In 2000 Pete had some health problems that impeded his farm activities. Among them was a disconnected shoulder cuff—he couldn't drive a tractor. About that time, he received a vision that led him to leave his farm and return with his wife, Mary Ann, to the Nisqually tribe. He sold the Vashon farm to Don Smith, a new farmer who named the enterprise Field Day Farm.

Pete became operations director for the Nisqually Indian Tribe in Yelm, helping them acquire land and also build community gardens. His health improved, and he healed his shoulder by working out at a local gym, he says. On his seventieth birthday, he climbed Mount St. Helens. He continues to plant trees on the property he bought with Mary Ann in Olympia, Washington. When asked how he's doing, Pete says, "Farming was a wonderful part of my life. I'm better off now, and Mary Ann is happy here. As long as I can be around plants, I have the spiritual benefit. I don't have to sell them to have that."

Field Day Farm (site formerly known as Pete's Farm), 23720 Dockton Road SW, (206) 463-9032

Part IV
COMMUNITY-SUPPORTED AGRICULTURE AND CONTEMPORARY FARMERS

Providing Healthy Food and Hope for a Future with Sustainable Agriculture

Chapter 10

Bob and Bonnie Gregson: Island Meadow Farm

Bob and Bonnie Gregson, Island Meadow Farm, retired from farming and created a program that keeps their land perpetually in agriculture. Photo © Kathleen Webster.

Island Meadow Farm is a charming, picturesque, vibrant few acres just off Cemetery Road, where vegetables, fruit, and salad greens are intensively cultivated on less than 1 acre. Head down the dirt driveway and park at their produce stand and you can sense the gentle, cooperative spirit of the land. Chickens cluck and forage on the other acres; nearby, pigs, llamas, and sheep graze contentedly, pausing to eye the visitors who have come to partake in the peacefulness here.

The little produce stand, with its entryway of climbing plants, houses a cooler that provides food for thirty to forty families who subscribe to twenty-two weeks a year of organically grown produce. Extra food, herbs, gourmet salad greens, flowers, and brightly jeweled herbal vinegars are displayed for walk-in customers seeking healthy, homegrown food.

It looks easy, but it wasn't always so. When Bob and Bonnie Gregson bought the 12.75-acre flower and nut tree farm on Vashon Island in 1988, they thought they would make their living with the resources left by the former owner. There were ten hoop houses filled with tulips, daffodils, irises, and freesias. There were nut trees that had been planted by the famed Shane brothers sixty to seventy years before. But the flower bulbs had botrytis, a permanent soil condition that mars the flowers, destroying their marketability. And the nut trees were at the end of their productive lives.

"If we could have lived on $1,000 to $2,000 a year, we could have survived," Bob jokes. Both in their forties, neither had a retirement income, though they had some savings. Neither of them had any experience farming, either. Bob had most recently worked in property management for Virginia Mason Hospital in Seattle after having a heat pump business and a military career, and possessed degrees in engineering and business. Bonnie had worked in clinic management at Virginia Mason, where the two met.

After much soul-searching, they decided to find a way to make the farm work. They planted vegetables and salad greens. They gutted and entirely remodeled the old house, and moved mountains of trash off the property. Bonnie was diligent in her upholstery business, and its income kept the two going for the first four or five years while they grew the farm business.

Theirs was a long piece of property, about 325 feet wide and 1,600 feet long. The house was at one end of it, at 10325 SW Cemetery Road. The first years, they lost a dozen sheep to dogs, because the home was too far away to maintain vigilance. They sold their home and its 2.75 acres and brought a prefabricated house to place on the middle of the remaining 10 acres, where they could reside closer to the crops and livestock.

Early on, they decided they wanted to focus on organically growing vegetables and other produce. About that time, islander Cindy Olson was looking for a new piece of land to lease to grow her salad greens, which were shipped around the country and earning $20 per pound. She leased .75 acres from the Gregsons, and they learned the business with her, then bought it from her when she moved off-island a year later. By that time, 1991, the Gregsons were making half their income from upholstery and half from farming. In 1997 Bonnie quit the upholstery business, because the eggs from 300 chickens provided their winter income.

Their biggest client, the Four Seasons Olympic Hotel in Seattle, had been buying 90 pounds of salad greens a week. In 1993, two weeks before their first delivery of the season, the bottom dropped out: The hotel chose to buy cheaper salad greens from California. "They said 'sorry,'" recalls Bob. "There we were with tons of salad capability. What were we going to do?"

They sought out Larry's Markets and the Thriftways on the mainland to sell their produce. Then Mark Musick, a long-time proponent of direct farm marketing, suggested, "You really ought to get into the subscription farming business. This is a thing of the future and you ought to try it."

Bonnie says, "At first we didn't want to. We didn't like the idea of accepting people's money up front without knowing whether we could produce the rest of the season. What if we got ourselves into a pickle and had to give people's money back? But we said we'd give it a try with twenty customers the first year. That was 1992. And it went really well."

Half of their customers were in Seattle, and Bob went to the mainland two days a week, delivering to florists and subscription customers. They also sold their produce through the then-recently established University District Farmers Market, as well as Seattle's Pike Place Market. They found the neighborhood farmers markets more lucrative, because they weren't competing with produce shipped in from wholesale sources and California.

But even the University District Farmers Market brought them only $300 in sales on Saturdays, so after a couple of years, the Gregsons dropped their mainland customers. They built a produce stand on their farm and continued to build relationships with their Vashon Island subscription customers, selling their excess produce at the Vashon Farmers Market. Eventually they found that their best profits were made by staying local.

"So now we don't go anywhere," says Bob. "All the customers come to us. Everything is retail. We don't wholesale. That's been the secret to success here. Not hiring outside labor, selling retail, and having good relationships with customers," he says. The Gregsons have hired apprentices to help them with the farmwork, and in exchange for their inexpensive labor provide the apprentices with mentorship, training, and housing.

After twelve years in business, the Gregsons wrote a book about their experience, titled *Rebirth of the Small Family Farm*. They self-published it, and it continues to sell well, mostly through seed catalogs. In the book, they share tips about the costs, challenges, and joys of making a living on just a couple acres of land. They say their farm income is about $40,000 a year, which can work for an individual or couple with a low mortgage payment.

Although many people are attracted to farming because it seems to offer independence and quiet solitude, Bob's involvement in agricultural politics has made a big difference for small organic enterprises such as Island Meadow Farm. He garnered effective support programs that advocate for sustainable agriculture and small organic farms by combining the political efforts, clout, and common goals of the Safe Food Initiative, the Washington Sustainable Food and Farming Network, and Puget Consumers Coop. Bob was the first organic farmer ever appointed to the Washington State University dean's advisory council. He also served as the chair of the King County Agriculture Commission from 1995 to 1998 and from 2000 to 2001. In addition, from 1998 to 2001, he served as acting chair for the Farm Link Program, which links people who want to get into farming with people who want to leave it. Through the consulting services of that program, Bob and Bonnie have designed a plan that will keep their land a farm in perpetuity so that they can retire. In their midfifties, their knees, wrists, and backs are giving out.

In February 2001 they sold the prefabricated home and farm business to Greg Kruse, a farmworker who apprenticed there for two years. The Gregsons now live in a two-bedroom home they built in 1999 on the same 10 acres, with a sunny workshop and greenhouse as its main floor. Through their contract with Kruse, the former apprentice pays the mortgage, the site remains in agriculture, and the Gregsons can farm for a hobby.

The Gregsons advise new farmers to start very small, with 1/2 acre of actual production land. They also say you should keep your day job for the first few years, preserve your body, stay involved with politics that support farming, develop a relationship with your customers, and do all the work yourself. The Gregsons say that farming has to be your passion, because when you meet challenges, you'll work them through. They advise new farmers to sell retail only, and have a value-added crop or product. In their case, their mixed salad greens are value-added because lettuce alone might sell for $1 a pound, but mixed greens that are packaged might sell for $10 a pound.

The Gregsons say they never used a formal business plan, but they used good business sense to know when and in what way to change course. They rejected the myth that a business has to expand in order to succeed. In the end, their approach to business is compatible with their approach to farming: "We just observe and interject ourselves minimally from time to time. A little tilling, a little planting, and a lot of harvesting," says Bob.

Greg Kruse is carrying forward their traditions, offering farm goods at the Island Meadow Farm produce stand to fifteen subscribers, the Vashon Farmers Market, and Wednesday's market at Minglement.

Island Meadow Farm, 10301 SW Cemetery Road, (206) 463-0053. Rebirth of the Small Family Farm is available at the King County Library, or by contacting Bob and Bonnie Gregson, IMF Associates, P.O. Box 2542, Vashon, WA 98070, (206) 463-9065.

Chapter 11

Ron Irvine: Vashon Winery

Ron Irvine has pursued a lifelong passion for wine and cider production. Photo © Kathleen Webster.

Ron Irvine never forgot the taste of the cider he had bought one day when, years ago, at the age of twenty, traveling in London, he stopped for a picnic. The scent of the cider was almost floral, like a baked apple coming out of the oven—he could smell cloves, yet it was sweeter, softer, rounder. It was almost like imagining a yellow-red rose scent, he says, recalling it more than two decades later.

"Ever since, I've been trying to figure out how to get that aroma," says Ron. "For me, it's my Holy Grail." His quest took him from England to France, to apple-growing centers in Mount Vernon, Eastern Washington, and Vashon Island, trying to duplicate that scent.

While the taste and scent of that cider remained in his soul, Ron worked in the wine and cider business for twenty-five years. In 1975 he and several partners founded the Pike & Western Wine Shop in Seattle's Pike Place Market. During the early 1980s he served on the Pike Place Market Historical Commission and the Pike Place Market Public Development Authority, and led movements to preserve the Pike Place Market and agriculture in the region.

When the shop first opened, there were only 3 wineries in Washington state and 3 in Oregon. Today there are 130 wineries in Washington and 140 in Oregon, Ron says. "I like to think that we helped, that we provided an avenue for selling the wines. We recognized their quality," Irvine says.

In the late 1970s, Ron moved to Vashon Island and commuted to the wine shop. Over the years, he continued his research about cider. In Europe, cider contains alcohol, Ron explains. In the United States, cider is nonalcoholic; we refer to alcoholic cider as "hard cider." Hard cider has 3 to 7 percent alcohol, compared with wine's 9 to 11 percent alcohol, Ron says. Cider apples are graded on their level of tannin and acid, and are classified into four types: bittersharp, bittersweet, sweet, or sharp. English ciders are dry. French ciders are sweeter, and their techniques are almost magical and artistic, says Ron.

As president of the Northwest Cider Society, and active in the Western Washington Tree Fruit Foundation, he continued his networking about cider production. In 1989 he borrowed space from the Vashon Winery and Wax Orchards Farms, and used apples from the Mount Vernon agricultural research station to manufacture a cider fashioned after the one he had tasted those years ago. He named it Centennial Cider to commemorate the 100th anniversary of the year the state's first apple tree had been planted, in Vancouver, Washington. Under his Irvine's Vintage Cider label, he manufactured varietal ciders from particular apples, such as the Cox's orange pippin and the Kingston black apple.

In 1991 Ron sold his share in the wine shop and began research on a book about winemaking in Washington state. During his research, he discovered that almost all the Vashon farms had at least one row of grapevines, of a variety called Island Belle, many of which had been established around the time of Prohibition in the 1920s. The first winery on Vashon was the Harley Hake winery in Dockton, established in 1936. Another was a blackberry winery on Burton Hill, a brand called Black Beret. The largest is the Tsgwale Vineyard, planted in 1924, which can be viewed while driving along Quartermaster Drive. Vashon Island currently has two wineries: Andrew Will Winery, as well as Vashon Winery.

While writing his book, Ron worked part-time managing the Vashon Winery for owners Will Gerrior and Karen Peterson, who had owned it since 1987. While there, he helped downsize the winery, carefully managing its inventory and sales of its cabernet sauvignon, merlot, and semillon. "It turned out to be an absolutely wonderful way for me to learn firsthand while I was writing my book about wine and wine history. It was also a therapeutic way, kind of an artistic way, and a physical way of relating to wine while I was sitting there working on a computer most of the day," Ron says. "I worked on the book in the morning until 11:00 and then I'd come over here and get on the forklift and move barrels around." In 1997 Ron self-published his book, *The Wine Project—Washington State's Winemaking History.*

In 1998 Ron assembled a chronology of apple research on Vashon and presented it to visitors to the Apple Bee celebration at Doug and Adrianna Havnaer's home, where they had planted an orchard of 1,500 apple trees in in 1992. Through that research, he discovered some interesting horticultural facts about Vashon Island. C. A. Tommison of Vashon established the first horticulture society in the state in 1892. Vashon settlers planted orchards on Vashon Island, among other places, to help establish their homesteading claim. Often men would arrive alone, establish their claim by planting fruit trees, and then go back and get their family or go off prospecting for gold, Ron says. Vashon's orchards looked so promising to state horticultural inspector Henry Huff that, in 1914, he was quoted as saying, "Vashon is the best fruit district in Western Washington."

In December 2001, Ron bought Vashon Winery and moved it to the cider barn of the Havnaer's farm, a mile north of downtown Vashon. As the new owner of Vashon Winery, Ron hopes to increase its production of wines and ciders produced from locally grown fruits. The yellow-green Chasselas grapes from the Tsgwale Vineyard are among those used in his wines.

Ron says he wasn't ever able to exactly duplicate that wonderful aroma of the cider he savored twenty-five years ago. But perhaps the journey has been more rewarding than reaching the goal.

Vashon Winery, 10317 SW 156th Street, (206) 567-0055, www.vashonwinery.com. The Wine Project—Washington State's Winemaking History can be obtained through local booksellers and by E-mailing sketchpub@aol.com.

Chapter 12

Jasper Koster and Martin Nyberg: Green Man Farm

Jasper Koster and Martin Nyberg and Jasper Koster of Green Man Farm are organic farmers and political activists who walk their talk about sustainable agriculture. Photo © Kathleen Webster.

In 1997, when Jasper Koster's employer offered her the choice of following the company's relocation to Richmond, Virginia, or accepting their generous severance package with retraining dollars, it took her all of a New York minute to think it over. Spend her days carrying out administrative tasks such as facilities management, lease facilitation, and contract negotiations, or make her living on the Vashon farm where she and her husband, Martin Nyberg, had just settled?

"After thinking about it for a nanosecond, I decided to take the retraining," recalls animated, enthusiastic Jasper. Starting a small farm venture had been a dream the couple already shared, and the opportunity to learn how to do it right was perfectly timed. She enrolled in South Seattle Community College's two-year environmental horticulture program in 1998, and halfway through the summer, the couple began their first growing season on the farm.

Right about that time, Jasper served on the Vashon Chamber of Commerce and met *Beachcomber* newspaper publisher Lee Ockinga. Jasper accepted Lee's invitation to write a column for the *Beachcomber* about the farmers of the Vashon Farmers Market. The column increased public awareness and appreciation of healthy, homegrown crops, gaining Jasper eight of the ten customers who subscribed to receive what she produced that season. Subscription farming is one of several CSA marketing methods in which the farmer contracts to grow crops for a number of customers who pre-finance the growing season. It benefits the farmer because the customer base is assured, and the customer is assured of receiving healthy produce on a regular basis directly from a known source.

The Nybergs named their farm the Green Man Farm, after the benevolent and kind Celtic deity of the plant kingdom. Martin, who hails from the mythical-poetic men's movement, particularly reveres the symbol because it represents the wildness and resurrection of nature, and the Green Man is one of the rare mythological male figures who doesn't have a tendency to destroy, he says. Martin and Jasper both play Celtic music as performers in the band Geordie's Byre. Jasper plays the harp, fiddle, and mandolin and sings, while Martin sings and plays guitar.

The Green Man Farm is a 2-acre plot on SW Dilworth Road, with just 1 acre under cultivation. Its bounty includes salad greens, brassica crops such as cabbage, broccoli, cauliflower, and kale; and root crops such as beets, carrots, parsnips, and potatoes. Also produced are snap peas, snow peas, pole beans, corn, winter squash, and summer squash. In their greenhouse, they grow peppers and tomatoes. Their fruit produce includes apples, sour cherries, plums, raspberries, Asian pears, Bartlett pears, strawberries, and golden plums. They have found that brassica crops grow particularly well in their farm's type of soil.

Also working the land alongside the Nybergs are forty-nine Araucana chickens with various monikers they earned by remarkable moments in their careers. There are roosters named Mick, Keith, Charlie (barnyard Rolling Stones), and Wanda ("a big, beautiful hunka chunka churnin', burnin'love!"), and hens named Poopey-Butt (don't ask) and C'mere, C'mere, C'mere. They chose Araucana chickens for their large pastel Easter eggs, and because they are good foragers who don't "bill out" when they feed. Billing out is a messy eating habit in which chickens toss feed around the yard, wasting it and attracting rats—"Ugh," says Jasper, declaring that the downside of farming. One morning's critter problem was finding that a deer had eaten much of the corn crop, she says.

Three sweet-tempered French Alpine goats complete the cast of characters: Leah; her daughter, Yuki; and an adoptee-orphan male named Merlin—all of whom are diligently chewing the blackberry bushes down to nubs. And of course, the smiling overseers: their pet dog, Wolfgang, the great fuzzball, and his buddy, Murphy.

Jasper grew up in a family that valued growing their own food. Raised in Southern California with vegetable gardens, fruit trees, and chickens, Jasper remembers that her mother always had a "bodacious vegetable garden and she composted everything imaginable." Her mother is Japanese, and her culture highly values growing, preserving, and conserving food, Jasper says.

"It's very common in downtown Tokyo; people are growing whatever they can in coffee cans and little tiny planters stuck out on windowsills. Even in that dense population, people really try to keep growing things."

During harvest season, Jasper cans and preserves produce into value-added products for her subscribers, including pickles, preserves, jams, jellies, salsa, and pesto. She gives these free to subscribers, and is researching the prospects for a time-share or community kitchen that could provide a cost-effective option for growers and food producers on Vashon Island.

Equally passionate and industrious about the value of growing food is her husband, Martin, who had been dreaming about farming since the mid-1970s. "The actual genesis started from reading *Mother Earth News,* About '5 acres and self-sufficiency,' and I just dreamed," Martin says. In the early 1990s, Martin was a caretaker for 35 acres in Enumclaw, and began studying the science of farming and how to start a CSA farm. He read a book called *Secrets of the Soil* that outlined some philosophies of farming, including Rudolph Steiner's biodynamic agriculture. That philosophy purports that food lost its vitality and ability to support spiritual evolution about 100 years ago, and that this trend can be reversed by utilizing certain homeopathic principles, enlivening the soil, and involving the community, Martin says.

Martin applied those principles to a compost heap experiment. He built a traditional rectangular one that was 6 feet high by 9 feet long. He built a second one that was shaped like a tetrahedron—a four-sided pyramid that is a sacred geometric structure, similar to chlorophyll molecules. He added certain biodynamic preparations to the points of the tetrahedron that corresponded to its internal organs, as if it were a living being parked head-first in the earth. Within twenty-four hours, the experimental compost pile heated up to 120 degrees Fahrenheit; within two weeks, it had reduced its size by about 20 percent. The traditional rectangular compost pile took six months to reduce its volume by half. Whether mystical, mysterious, or just plain practical, such methods helped shorten the time it took to build tilth on their farmland, where their soil was heavy with clay.

Martin can sum up the downside of farming for him in one word: "Commuting. When we made the decision that one of us needs to stay here and one of us needs to go to work to help pay off the mortgage, it was a little bit like a funeral for me. Making that decision was hard. I feel bad every time I pull away and get on a ferry to Seattle," he says. Martin commutes to Renton five days a week, where he works in high-tech sales and marketing. Jasper works part-time as a bookkeeper for Vashon HouseHold, an organization that helps provide affordable housing.

Is farming, for them, an economically viable venture? "It's really difficult," says Jasper. "The cost of living on Vashon is higher than in many places around Seattle. Property is so expensive here now. Our mortgage is high. Rents have gotten out of control. The way to make this economically viable is to not have a mortgage. And even if you buy some crude land, water's expensive; so are the septic systems. It's harder and harder for people to get access to the land," she says.

After their third growing season, they calculated it would take ninety subscription customers for them to gross $40,000 a year, counting supplemental income from their produce stand and the Vashon Farmers Market. After costs and mortgage, that would net only $10,000. They are looking into alternatives, such as leasing the neighboring acreage through King County's Farm Link Program, or asking subscribers to provide workshares.

Ever busy and industrious, the duo stay involved in several E-mail and discussion groups that help promote the cause of healthy food and farming. Martin established an on-line fresh food exchange for people who want to sell or trade their excess crops. Jasper wants to write a resource guide for commercial as well as hobby growers on Vashon. Both have served as co-presidents of the Vashon Island Growers Association. In May 2002, Martin was appointed to serve a two-year term on the King County Agriculture Commission, and vows to advocate for the protection and promotion of sustainable agriculture.

Both Jasper and Martin remain motivated to keep the farm going, because they are concerned about the health consequences of much of the food that is consumed in the United States.

"I'd like to dispel the myth that it's OK to buy all your food from the grocery stores that import 90 percent of their produce from California and Mexico without expecting major health consequences," says Martin. "Ninety percent of the dioxins, the chemical pollutants found in every human being's body, come from food sources. It's a major problem in men's fertility rates and cancer. Sixty percent of processed foods are guaranteed to be genetically modified in some way. In the United States, four transnational corporations control a majority of the food sources: beef, 79 percent; chicken meat, 50 percent; pork, 50 percent; and flour, 62 percent. Globally, two corporations control 75 percent of all grain products," he continues.

He advocates that Vashon Islanders speak out against such practices and support the farmers on Vashon by subscribing and buying their crops, or by giving up a few years of lawn-mowing by leasing that land to a grower in the name of health and community instead of profit. Martin also urges islanders to use their powerful voice through the Internet, as a weapon to advance the cause of small-scale sustainable agriculture.

Says Martin: "We just may be preventing the next famine."

Green Man Farm, 8800 SW Dilworth Road, (206) 567-4548, www.GreenManFarm.com

Chapter 13
Amy and Joseph Bogard: Hogsback Farm

Amy and Joseph Bogard, and daughter, Liesl, Hogsback Farm, bought land formerly owned by strawberry farmer Tok Otsuka, where migrant farmworker cabins (in background) still stand. Photo © Kathleen Webster.

In the early days of Hogsback Farm, a group of middle school kids arrived for a tour. When they stepped out onto the field, the first thing they said was, "Can we have some plastic bags to tie around our shoes so they won't get dirty?" Farmers Amy and Joseph Bogard humored them, and regaled them with stories about how good their homegrown carrots and scrambled eggs were going to taste later on that day.

"Well, we don't like carrots," one said.

During their visit, the students toured the farm, learned about organic farming, learned what community-supported agriculture means, then rolled up their sleeves and went to work—and began to feel they were part of something grand. Later Amy made them scrambled eggs from the farm's chickens. One of the kids said, "These are the best eggs I've ever had."

"The carrots, too," another said.

"It was amazing to see that transformation over the day," recalls Amy. "And those kids did unbelievable work! They deadheaded all the flowers along our fence. They pulled a 400-foot row of broccoli. They cleaned out a salad bed, and then they picked the rest of our carrots and cleaned them all. It was amazing to see how those kids converted in a day into just really being *into* being here. By the end of the day, they didn't even care whether they got dirty or not," she says. Afterward, the students sent a warm thank-you note.

Amy and Joseph Bogard hope to increase their farm tour–work experience opportunities for youth, because they both really enjoy combining farming, teaching, and sharing their ecological philosophies. They are modern-day homesteaders who arrived in 1996 and 1997, respectively, after buying the land for their farm with a couple of friends. For the first years, they lived in tents and in an old, unheated migrant farmworker's cabin on the property. They built a chicken coop and a crop processing building before constructing their house. Amy jokes good-naturedly that "our chickens had a condominium before we had a place to live."

Like other contemporary Vashon farmers, the two shared the farm dream for a time while they extracted themselves from their former lives. Amy was a mechanical engineer turned ski bum turned school teacher who hailed from Maine and ended up in Seattle. She taught in Seattle schools for five years, at Washington Middle School and West Seattle High School. During that time, she served with Americorps, a national educational organization that helps youth learn environmentally based leadership skills.

Joseph is from Southern California. He attended agricultural school at University of California Davis, worked on an organic farm in Oregon, and farmed a P-patch in Seattle. He still works three-quarter time at a salmon-advocacy coalition called Save Our Salmon, headquartered in Seattle. Amy works full-time on the farm.

Visitors to Hogsback Farm often ask Amy and Joseph, "Where are the hogs?" Answer: There are none. Hogsback is derived from the geological term "hogback," for a ridge of land that is precipitous on either side. Hogsback Farm is 12 acres, 2.5 acres of which are fenced off, that formerly was a strawberry field owned by Tok Otsuka, one of Vashon's Japanese-American farmers (see Part II, *Strawberry Fields Forever*).

When the Bogards first tested their soil, they found it was almost totally devoid of nutrients. Undaunted, they brought in fifty truckloads of horse manure and mostly hand-spread it, then planted six successive cover crops to add fertility and tilth to the soil. They hired longtime Vashon Island tractor worker Gil

Hartness to help spread the seeds for some of the cover crops. Joseph remembers coming home at night to see a headlight bouncing around in the field in the dark, coming toward him. "Amy was in the bucket, throwing seed in so Gil could come back behind it and run it into the soil."

Says Amy: "It was a full moon and it was absolutely gorgeous."

Amy's skills as a carpenter, electrician, and plumber helped save them money while they built the physical infrastructure of the farm. To learn about farming, they both talked with other farmers on the island to find out what worked and what didn't. They got involved with the Vashon Island Growers Association and in 1998 attended the annual Seattle Tilth Conference. They visited subscription farms in the region. That year, they grew enough food to provide excess crops for friends and for their wedding party that August.

By year-end 2000, the Bogards were living in a heated space with hot and cold running water, which made it easier to raise their baby, Liesl, who was by then nine months old. Half their farm income was from their twenty-five to thirty-five subscribing customers, with the remainder of their income from sales at their on-site produce stand and the Vashon Farmers Market.

Today they have an orchard of twenty-five producing trees. They have eighty chickens, including barred rock, black and golden sex links. They also have a U-pick flower garden. Their crops include tree fruits, salad greens, herbs, cooking greens (including chard, mustards, and kales), potatoes (including yellow finns, russets, and Yukon golds), root crops (such as turnips and parsnips), as well as cabbages. Their farm infrastructure completed, they netted $8,000 in 2000.

What is the downside of their style of farming? "The killer commute," says Joseph. "And the work is relentless. You never really feel you can sit down in a big, soft chair for very long before you have to get up again." Says Amy, "The weeds. The stress of getting things in the ground on time. I think in time we'll become more seasoned, more able to roll with the punches, more established, more organized, more strategic. But otherwise it's good. It's satisfying."

Hogsback Farm, 16530 91st Avenue, (206) 463-1896

Chapter 14

Margaret Hoeffel with Jean Bosch: Blue Moon Farm

Margaret Hoeffel and her husband Martin Koenig of Blue Moon Farm. Margaret partners with Jean Bosch growing greens and other crops at a cohousing site on Vashon. Photo © Kathleen Webster.

Blue Moon Farm was a dream nurtured by two people looking for a way out of the intensity of the New York performing arts scene. Margaret Hoeffel, a dancer of twenty years—which she pegs as a long diversion from her true calling—always wanted a market farm. She was a longtime member of a food co-op; her mother's family history was one of peasant life. Her mother harvested, canned, and froze the produce from their backyard. Margaret knew the sound of a strawberry coming away from the plant; it's a sound in her genetic code.

"There I was living in New York City, doing this entirely other thing," Margaret recalls. She felt her dancing years winding down and an urge to start a family, and transitioned into garden work, a project that combines creativity, art, and botanicals. "In an urban setting, I was doing whatever I could to get my hands in the dirt. I always wanted to be a farmer."

In 1984 she and her husband, Martin Koenig (see Chapter 15), visited Seattle and Vashon, looked for housing, and learned of the cohousing project on Vashon's Bank Road. They joined, and four years later they moved in. Margaret began planting successions of buckwheat and winter rye to lend fertility to the land. The following year she built raised beds and planted three beds of winter greens, mostly kales and salad greens. At first it was a community garden of several neighbors. "And then it was, well, me," Margaret says.

From 1995 to 1999, Margaret worked part-time for the Vashon–Maury Island Land Trust while she transitioned into farming. During her tenure she teamed with the Vashon Island Growers Association and organized a series of workshops about farming on Vashon, called "Get Growing" and "Tales from the Tractor." Those workshops brought together former and current farmers, who swapped tales about the internment of Japanese farmers, the challenges of a limited labor pool, the capriciousness of the markets, and the thrills and satisfaction of a simple, sustainable life on the farm. (Those stories are available on videotape, and those events led to the creation of this book.)

The workshops carried forth part of the land trust's mission to preserve the rural character of Vashon Island, but applying it to acquiring and preserving land can present problems, Margaret says. "It's difficult because when you preserve farmland, you need to have someone who is farming. You can't just preserve it because it was once farmland. It doesn't grow corn by itself like a forest grows trees. We need to have farmers," she says. In 1999, the land trust hosted a community meeting to assess the possibility of acquiring Pete Svinth's farm when he was leaving Vashon (see Chapter 9). Those efforts didn't bear fruit, but instead, the land was sold to another farmer.

Margaret partnered with Jean Bosch, who was just ending her tenure with Vashon HouseHold, and in 2000 they created Blue Moon Farm. The farm is named for fertility and new life, because Margaret believes her first child was conceived on a blue moon—the second full moon within a calendar month. Early on, the Blue Moon Farm proprietors saw winter vegetables as their niche, and envisioned a subscription CSA farm. Later, they planted strawberries, blueberries, and raspberries, then added chickens to their inventory. By the year 2002, the two probably had 1 acre in cultivation, with thirty-seven raised beds that are 200 square feet each—including three hoop greenhouses, totalling 10,000 square feet. In the fields, they grow winter squash, potatoes, and other root crops, and the fields are rotated with cover crops to build the soil's fertility and tilth. The marketing plan is to provide salad mix from spring through fall, skipping August; then they add berries and root crops toward the fall. During the winter, they teach creative ways with parsnips, rutabagas, and parsnips through three cooking classes offered to their subscribing customers. In addition, they sell produce at the Vashon Farmers Market in downtown Vashon, as well as to Minglement.

Margaret currently works thirty hours a week on the farm, balancing her efforts with raising the couple's three children, Ravenna, who is twelve, and twins Ezra and Avalon, eight years old. Ravenna has been apprenticed to help harvest edible flowers to enhance the salad greens.

Jean works full-time on the farm. "It's a seven-day week. The plants don't wait for you until Monday. But it's great to be outdoors every day, great to hear the birds, and see the sun," Jean says. "And the Vashon community is amazingly supportive. If we could produce more, we could sell more."

Is the farm is economically viable, enough to sustain a family and a mortgage payment? "Not on this scale," Margaret replies. "I think you need to be bigger, and you need to be mechanized. You need to have maybe 5 acres." She refers to discussions with the National Market Farming listserve, and a trend toward larger mechanized farms of 11 or more acres. Blue Moon Farm currently serves twenty-five to forty subscribing customers. One of Margaret's marketing ideas is to have a little farmers market at the grange building near Vashon's north end ferry dock, to sell wholesome food to commuters.

Currently, Blue Moon Farm is one of six farms leased on the adjacent 4-1/2 acres of a 19-acre parcel called Roseballen Community Land Trust, named for Rose Ballen, longtime Vashon resident, one of the founders of Vashon HouseHold, and a proponent of cooperative housing.

Given the hard work and little profit, why does Margaret want to continue farming? She recalls a presentation she attended at which the speaker raised the concern that our country is going to lose its soul because we can't see where our food is growing. When you eat from the land around you, you become a part of it in a way.

"There's a lack of connectedness in our society, a breakdown of our world culture, that needs to be reconnected by what we eat from our land," Margaret says. "I feel we have a unique opportunity as an island community because we are isolated in one way, and we are connected to each other in another way."

Blue Moon Farm, 10421 SW Bank Road, # 18, (206) 463-1224

Chapter 15

Martin Koenig: The Raising of Village Green

Today, Village Green, in the center of downtown Vashon, is the home of the Vashon Farmers Market, and a favorite Saturday destination for consumers who want to buy produce directly from local farmers.
Photo © by Ray Pfortner.

Martin Koenig will tell you he isn't a fund-raiser, yet he was able to raise more than $250,000 in a matter of weeks to help acquire a choice piece of real estate in the heart of downtown Vashon. Today, dubbed Village Green, it belongs to the citizens, through the Vashon Park District. Among other uses, it remains the home of the Vashon Farmers Market.

The little plot of green land across from Bob's Bakery had been owned by three people: Vashon Realtor Mary Jane Brown, Larry Baxter, and Gilbert Chase, the latter two partners who had since moved off-island. Brown had long welcomed farmers on this land, initially giving them free use of it for more than ten years, since the Vashon Farmers Market's early days.

When Vashon Island Growers Association (VIGA) was in its infancy, volunteer farmers gathered materials and donations and erected a wooden structure on Brown's land from which to sell produce. During the winter, crafters set up a big, square, circus-like tent for selling their wares. In the summer the site attracted live musicians, and the customer base grew. Folks knew they could purchase healthy produce and nursery plants, and meet the people who grew them. Going to the Vashon Farmers Market and Bob's Bakery became a regular Saturday ritual for many Vashon citizens.

Martin Koenig had worked twenty-five years in the arts field in New York City. He founded and directed a 501(c)(3) arts organization in New York that presented traditional dance events for communities on about $500,000 a year. He also worked in the Smithsonian Institution for the Festival of American Folklife. In 1984 Martin and his wife, Margaret Hoeffel (see Chapter 14), left New York and relocated to Vashon Island.

Martin recalls the day in the early spring of 1999 when he received a phone call from friend and realtor Laura Wishik, who was a proponent of the preservation of farmland. She told him that the little plot of land across from Bob's Bakery was going on the open market that coming Saturday. Laura knew Martin's and Margaret's philosophy of farm advocacy, and urged him to get in touch with VIGA.

Concerned that the Vashon Farmers Market could lose its gathering place, he called Marie Harrington, then president of VIGA, and said, "Marie, I just got a phone call. You've got a week to raise $200,000." He let it go with that for a couple of days and, seeing that nobody was taking any action, that Wednesday called her back and offered to raise the money.

Martin reported to VIGA members and said, "Look, I'm not a fund-raiser, but I'm enthusiastic about this. I'm passionate about this. I think that this is good. But I'm not sure who should be taking the leadership on this." He presented the vision to the Vashon–Maury Island Land Trust, but found that their energy and resources had recently been drained by the completion of another fund-raising campaign.

By then a chapter of Washington Tilth, VIGA was a 501(c)(3) nonprofit, which qualified it to receive funds that donors could deduct from their income taxes. But VIGA, as a group, focused on cooperative farming and marketing, and had never faced a project of this scale before. Martin asked Marie, "Do you mind if I go out and try to raise this money?" He got a green light.

Martin got on the phone and spoke with prospective donors, who agreed to loan money at a low interest rate, with the property being the collateral. The first two donors, who were anonymous, committed $50,000 each. The *Vashon Beachcomber* newspaper carried a story about it, complete with banner headline, and public enthusiasm grew.

In a matter of nine weeks, Martin was able to raise $325,000 and earned a fee for his efforts. The money was raised by September 1, 1999, and the ownership immediately passed from VIGA to the Vashon Park District. Park District ownership will provide stewardship and multiple community uses, including the Vashon Farmers Market, assures Martin. Community members have since discussed and designed plans for the Village Green, which include its uses, landscaping, and maintenance.

What is the probable future of Vashon farmers and the Village Green? Martin says, "I see Village Green as being the heart of this community. It will give a prominence to VIGA and a prominence to the Vashon Farmers Market that really hasn't been there before. From all the press it has received—even more than raising the money to buy the property—islanders are educated about food, the farmers market, and VIGA.

"And I'm very pleased about that. The timing was right—because you have a whole new group of people coming to the island who are very serious farmers, and it's remarkable."

Vashon Island Growers Association (VIGA), P.O. Box 2894, Vashon WA 98070, (206) 567-4548, E-mail VIGA@GreenManFarm.com

Afterword

To assure a steady supply of locally grown, healthy produce, support your local farmer. Buy direct, become a farm subscriber, lend your land for crop production.

Appendix

Vashon–Maury Island Farming, Historical, and Land Preservation Organizations

Mukai Farm and Garden, 18017 107th Avenue SW, Vashon, WA 98070. Call Jon Knudson for information at (206) 463-6935. Its founding organization, Island Landmarks, researched and designated historical landmarks around Vashon Island, including the Beall Greenhouse Company and the Mukai Farm and Garden.

Vashon Island Growers Association (VIGA), P.O. Box 2894, Vashon WA 98070, (206) 567-4548, E-mail VIGA@GreenMan Farm.com.

Vashon–Maury Island Heritage Association, P.O. Box 723, Vashon, WA 98070, (206) 463-7808. The Heritage Association has historical artifacts, photographs, and displays, and hopes to have an open-to-the-public museum by the year 2003.

Vashon–Maury Island Land Trust, P.O. Box 2031, Vashon, WA 98070, (206) 463-2644.

INDEX

Alien Land Law (1921), 56, 68
American Orchid Society, 37
Andersen, Conrad, 20
apple: Bee; pomace; varieties: Cox's orange pippin, Gravenstein, Kingston black, 104, 105
Augie's U-Cut, 50
Ballen, Rose, 122
Bank Road, 5, 43, 50, 76, 120, 123
Baxter, Larry, 125
Beachcomber see Vashon Beachcomber, 126
Beall Pedigreed White Leghorn Farm, 28, 36
Beall Road, 11
Beall, Allen; Cass; Chris, Connie; David, Ferguson, L.C.; Sarah, Tom; Tom Jr; Wallace, 25-38
Beardsley, Warren, 20
berries: blackcap raspberries; blueberries; boysenberries; currants; gooseberries; loganberries; Olympic berries; raspberries; see also strawberries, 4-7, 11-13, 15-24, 39-76, 82, 87, 109, 121, 122
Big Bear grocery stores, 47
biodynamic agriculture, 111
Bob's Bakery, 90, 126.
Bogard, Liesl, 114, 117
Bone Factory see also Pacific Research, 4, 33
Borden's, 57
Brink, John, 10
Brown, Mary Jane, 125
Bruner, 21
Brunson, Wallace, 10
Burton, 11, 56, 104
Burton Hill, 104
cane berries see berries
Cemetery Road, 4-5, 8, 96-97, 101
Center, 35, 63
Chase, Gilbert, 125
cherries; varieties: Stella, 11, 23, 55, 80
Cherrywood Station, 82
chickens; varieties: Araucana, barred rock, black, golden sex

links, white leghorn; see also poultry, 5, 10, 17-21, 80-82, 88, 96-97, 109, 110, 115-117, 121
Chinese Exclusion Act (1882), 67
Christensen, Conrad; Margareta, 19, 21
churches: Catholic; Lutheran; Methodist; St. John Vianney Church, 4, 5, 21, 61
Cisco, 10
Cletrac aka Cleveland Tractor Company, 23
Clough, Bill, 10
Coastal Salish Culture, xxiv
cohousing, 119-120
College of Puget Sound, 59
Colman Dock, 45
Colvos; dock; Passage; store, 14, 17, 18, 19, 24, 43
Committee to Preserve the Mukai Farm and Garden, xiii
community-supported agriculture (CSA), x, 13, 108, 110, 115, 121
Cove; dock; Road, 5, 14, 17, 20, 21, 36, 73, 82
cover crops: buckwheat, winter rye, 116, 117, 121
Cuttler, Reg, 10
Dairy Queen, 5
Dallas Times Herald, 84
Dannevig, 20
Depression see Great Depression, xxiv, 55
Diehl, "Dutch", 74
Dilworth Road, 109, 113
Dockton, 92, 104
Dumbleton, 10
Dunn, Tom, 10
Dunsford, Morris, 75
Eastly, Doc, 24
Edwards, Jacob, 20
Ellingsen, Andrew, 19
Ellingsens, 18
England and Peterson, 10
Erickson, Pete, 11
Ernisee, Fred, 7
Evans, Nancy, 31
exotic-animal ventures: chinchillas; goldfish; mink; nutrias; ostriches, 11
extension service: county, state, 89
Fagan, 10
Farm Link Program, 99, 112
Federal Bureau of Investigation (FBI), 62
Field Day Farm, 92
fishing, 1, 14
flowers: camellias; carnations; croft lilies; daffodils; dahlias; Easter lilies; gardenias; orchids; roses,

11, 27, 29, 30, 31, 32, 36, 37, 38, 96, 115, 122
Fujioka, Tashio, 6
Ganley, John, 10
Gerrior, Will, 104
ginseng, 11
glacial till, 81
Gonzaga University, 88
Gorsuch area, 43
grafting, 89-90, 139
grapes; varieties: Chasselas, Island Belle, 11, 18, 57, 59, 104, 105
Great Depression, xxiv
greenhouses, 25, 26, 27, 29-37, 121
Gruenewald, Charles, 59
Gruenewald, Mary (Matsuda), x, xiii, 53-68
H. Harrington Company Greenhouses, 34
Havnaer, Adrianna; Doug, 105
Hamer, Hans and Maren, 16
Hansens, 20
Harrington, Ella; Hilen; Jessie; Marie, 34-38, 126
Hartness, Gil, 117
Havnaer, Adrianna; Doug, 105
hay, 4, 7, 12, 18, 20, 55, 59
Hearst, Seymour, 10
Heights' water tank, 4
herbs, 96, 117

Hoeffel-Koenig, Avalon; Ezra; Ravenna, 119-120, 122, 124, 125
Hooper, Charles; Nancy, 38
hops, 82
Horiye, Mitsuno, 55
Houghton, Louis, 10
Huff, Henry, 105
Hunts' Brothers, 57
Huseby, Bernt, Ragna, 19
internment, x, xiv, 6, 45, 51, 59, 60, 66, 68, 75, 121
Island Landmarks, 131
Issei, 55, 71
J. H. Smucker Company see Smucker Company, 74
Jack's corner, 4
Japanese Americans, 44, 50, 51, 56, 59, 60, 61, 62, 67-68, 71, 75, 116
Japanese Association, 61
Jenn, George, 11
Jenson, Andy, 10
Johansen, August; Barton, 15, 18, 21
John L. Sexton Company see Sexton Company, 74
Johnsen, Aasman, 20
King County: Agriculture Commission; Farmlands Preservation Program; Health Depart-ment;

Landmarks and Heritage Program, xi, xiii, 48, 84, 99, 112
Kneebone, Frank, 10
Kresses, 20
Krockset, Ivor; Johanna, 17
Kruse, Greg, 100
K-2, 12, 31, 56
Kvisviks, 21
Lande, Ed, 10
Langsness, Ed, 19
Larry's Markets, 98
Larsen, Leo, 15
Larson, Kenny, 12
lead arsenate, 23
Lelands, 18
Lewis, A. J., 10
livestock: beef cattle; dairy cows; goats; hogs; llamas; sheep, 5, 10, 11, 88, 110, 96, 97, 110, 113
logging, xxiv
Lokke, Andrew, 19-20
Lorentzen, Andreas; Arthur (Tom); Bea; Ingvald, Leonard John, 14-24
Lucky grocery stores, 47
Mace, Floyd, 10
Mackie, Charlie, 82
Magnolia Beach, 56
Mann, Harold; Karl; Rosie, William, 3-8, 49
Marpole Culture, xxiv

Maryland Agricultural College, 35
master gardener, 89
Matsuda, Heisuke; Mary (Matsuda) Gruenewald; Mitsuno; Miyoko; Yoneichi, 7, 53-56, 58, 59, 60, 61, 62
Matsumoto, Frank; Jimmy, 6
McCarran-Walter Act (1952), 68
McMurray Middle School, 46
migrant workers, 46, 48
Minglement, 100, 122
Misty Isle Farms, 84
Mitchell Tree Farm, 11
Miyoshi, 12
Molviks, 18
Mooney, Ed, 10-11
Morgan Hill, 4
Morgan, Elicia, 10
Mosquito Fleet, 43
Mukai Cold Process Fruit Barrelling Plant, 75
Mukai, B.D.; Kuni, Masahiro, 17, 21, 22, 68, 71, 72, 74
Musick, Mark, 98
Nakamichi, 11
Nakanishi, Sato, 72
National Fruit Cannery, 46
National Market Farming, 122
National Register of Historic Places, 71

Native Americans, 12, 47, 57, 82, 91
Nisei, 45, 51, 55, 66, 75
Nishiori, 10
Nisqually Indian tribe, 87, 92
Northwest Berry Packers, 47
Northwest Cider Society, 104
oats, 12
Ockinga, Lee, 108
Olsen, Carl; George; John, 18, 50
Olson, Cindy, 97
orchards see tree fruits; trees, x, 80
organic farming, 115
Orient, the, 19
Otsuka, Tok, 6-7, 65, 76, 114, 116
Owen's Antiques, 6
Pacific Fruit and Produce, 22
Pacific Research see also Bone Factory, 4, 31, 33
Paradise Valley, 10, 12
Pascov, 10
Peterson, Karen, 105
Pike and Western Wine Shop, 103
Pike Place Market; Historical Commission; Public Development Authority, 98, 103
Pinedale Assembly Center, California, 65
Pioneer, 43
Portage, 43

poultry see also chickens, xxiv, 11, 28, 36
Puget Consumers Co-op, 99
Puget Sound Indian War (1854), xxiii
Puyallup and Sumner Berry Growers Association, 74
Quartermaster Drive, 10, 13, 104
Reine, Mrs, 19, 21
Rocknesses, 19
Roen, Al, 20
Roseballen Community Land Trust, 122
Rose, Clarence, 10
rural character, definition of, 121
Ryan, 43
Safe Food Initiative, 99
Sakahara, 22
Sakamoto, Sam, 11
Sansei, 55
Seattle Tilth Conference, 117
Seigrist, Frank, 10
Sestrap, August; Betsy; Kathryn; Robert, x, 22, 79, 82, 84
Sexton Company, John L., 74
Sextons, 20
Shane brothers, 88, 96
Shattuck, 22
Sheffield, George, 11
Sherman, Francis; Roger, 10, 50

Shinglemill Creek, 21
S'Homamish people, xxiv
Smith, Don; Will, 92
Smucker Company, J. H., 74
Soper Road, 29
South Seattle Community College, 23, 108
Spokane Street Cold Storage, 74
Steen mill, 21
Steiner, Rudolph, 111
Stewart, Tom, 84
strawberries: farms; picking; weevil; varieties: Magoon, Marshall, 39-76
Stopper, a horse, 14
subscription farming, 98
Sundberg, Albin, 21
sustainable agriculture, 93, 99, 107, 112-113
Svinth, Mary Ann, 92
Swain, Anna Sestrap, 84
Tacobet, 91
Tahlequah, 10-11, 48, 80
Takatsuka, August ("Augie"); Aya Taki; Yohio (George), 6, 7, 24, 41-50, 65, 76
Tanaka, 10
Taylor brothers, 73
Televiks, 20
Thomsens, 18

Thriftway shopping center, 30
Tjomsland, 21
Tommison, C. A., 105
Tree Fruit Association, 83
tree fruits: figs; peaches; pears; prunes; see also apples; cherries, 11-14, 19, 20, 22, 23, 55-56, 59, 79, 80-83, 89, 104, 109, 117
trees: American linden; basswood; Chinese chestnut; Christmas trees; Douglas fir; filbert; holly; oak, 11, 13, 30, 41, 49, 50, 74, 76, 88
Trones, 20
Tsgwale Vineyard, 104-105
Tule Lake, California, 45, 52, 65
University District Farmers Market, 98
University of California Davis, 116
U-pick, 50, 117
U.S. Department of Agriculture, 74
Usui, Mr., 61
Valley Packers, 47
Vashon: town; Auto Freight; Chamber of Commerce; dock; Farmers Market; Glacier; Highway; HouseHold; Landing; Park District; Pharmacy; Post Office, xxiii, 5, 10, 29, 30, 44, 57, 63, 76, 90, 98, 100, 108, 112, 117, 121, 122, 123, 124, 125, 126, 127

Vashon Island: Golf and Country Club; Growers Association (VIGA); Packing Company (VIPCO); Strawberry Festival, 10, 21, 22, 26, 38, 41, 47, 50, 63, 69-76, 82, 89, 104, 112, 117, 120, 125, 127, 131, 143

Vashon-Maury Island: Heritage Association; Land Trust, vii-xiii, 131

vegetables: asparagus; cabbages; cucumbers, peas; potatoes; radishes; rhubarb; tomatoes, 11, 12, 13, 17, 19, 24, 29, 34, 55, 56, 59, 64, 96, 97, 109, 117, 121

Vermuelen, Don, 10

Village Green, 90, 124-125, 127

Vye, Fred, 11

Walls, Lawrence, 18

Walter Bowen Company, 72

Wangs, 20

Warren, Dave, ix

Washington State University, 23, 99

Washington Sustainable Food and Farming Network, 99

Wax, August; Johanna; Linda; see also Sestrap, 80, 82, 84

Western Washington Tree Fruit Foundation, 104

Westside Highway, 17

Whitman-Beardsley place, 20

Wholesale Row, 72

wineries: Andrew Will Winery; Black Beret blackberry winery; Harley Hake winery; Vashon Winery, 102, 104, 105, 106

wines: cabernet sauvignon, merlot, Semillon, 105

Wise, Royce, 5

Wishik, Laura, 126

World War I, 36

World War II, xx, xxiv, 6, 22, 30, 36-37, 44, 46, 50, 52, 58-60, 67-68

wreaths, 49-50

Yakima Valley College, 31

Yoshidas, 65

Zarth, 21, 24

CONTRIBUTORS

Dr. Bruce Haulman is a twenty-eight-year resident of Vashon, teaches history at Green River Community College, and is working on a new history of Vashon-Maury Island titled Vashon: The Natural and Human History of an Island. *He also developed and manages the web site www.vashonhistory.com.*

Mary J. Matthews is a consultant in historic preservation. She helped organize a historic preservation movement on Vashon Island, starting in 1991, and was the director of Island Landmarks until 2001. She is currently a volunteer for the Mukai Farm and Garden.

Ray Pfortner is a photographer who specializes in landscape and environmental issue photography. He lives on Vashon Island.

Kathleen Webster is a photographer who specializes in natural outdoor portraits, hand-tinted and stock photography. She lives on Vashon Island.

Digital photo restoration of cover photo is a collaborative effort of Imago Imaging, Seattle. Imago Imaging specializes in high resolution drum scanning and digital preservation of photographic images.

About the Author

Pamela J. Woodroffe writes about farming, gardening, women's health, and family, among other topics. She is a former farmer of Sweet Woodroffe Herb Farm of Vashon Island, a farm that celebrated seasonal festivals, presented workshops about herbal history and folklore, and provided herbs and food for customers nationwide. She is also one of the founding members of the Vashon Island Growers Association. Before farming, she worked as a journalist and magazine writer, and has facilitated writers' support groups. A descendant of Nebraska homesteaders and a seven-generation farm in Norway, she lives in Seattle, writes for a living, and gardens for a hobby.

0-595-24133-6